A CATALOG OF
SPECIAL PLANE CURVES

A CATALOG OF SPECIAL PLANE CURVES

BY J. DENNIS LAWRENCE

Dover Publications, Inc.
New York

Copyright © 1972 by Dover Publications, Inc.
All rights reserved under Pan American and International Copyright Conventions.

Published in Canada by General Publishing Company, Ltd., 30 Lesmill Road, Don Mills, Toronto, Ontario.
Published in the United Kingdom by Constable and Company, Ltd., 10 Orange Street, London WC 2.

A Catalog of Special Plane Curves is a new work, first published by Dover Publications, Inc. in 1972.

International Standard Book Number: 0-486-60288-5
Library of Congress Catalog Card Number: 72-80280

Manufactured in the United States of America
Dover Publications, Inc.
180 Varick Street
New York, N.Y. 10014

CONTENTS

	Page
Abstract	viii
Introduction	ix
Notation	x
1. Properties of Curves	1
1.1. Coordinate Systems	1
1.2. Angles	7
1.3. Changes Between Coordinate Systems	12
1.4. Changes Within Coordinate Systems	16
1.5. Distances	18
1.6. Curve	20
1.7. Curvature	21
1.8. Mensuration	26
1.9. Geometry	28
2. Types of Derived Curves	39
2.1. Evolute, Involute, and Radial	40
2.2. Parallel Curves	42
2.3. Inversion	43
2.4. Pedal Curves	46
2.5. Conchoid	49
2.6. Strophoid	51
2.7. Cissoid	53
2.8. Roulette	56
2.9. Isoptic	58
2.10. Caustic	60
3. Conics and Polynomials	61
3.1. Conics	62
3.2. Circle	65
3.3. Parabola	67
3.4. Ellipse	72
3.5. Hyperbola	79
3.6. Power Function	83
3.7. Polynomial	84

		Page
4.	Cubic Curves	85
	4.1. Semi-Cubical Parabola	85
	4.2. Tschirnhausen's Cubic	88
	4.3. Witch of Agnesi	90
	4.4. Pedal of a Parabola	94
	4.5. Cissoid of Diocles	98
	4.6. Right Strophoid	100
	4.7. Trisectrix of Maclaurin	104
	4.8. Folium of Descartes	106
	4.9. Trident of Newton	110
	4.10. Serpentine	111
5.	Quartic Curves	113
	5.1. Limacon of Pascal	113
	5.2. Cardioid	118
	5.3. Lemniscate of Bernoulli	121
	5.4. Eight Curve	124
	5.5. Bullet Nose	128
	5.6. Cross Curve	130
	5.7. Deltoid	131
	5.8. Conchoid of Nicomedes	137
	5.9. Kappa Curve	139
	5.10. Kampyle of Eudoxus	141
	5.11. Hippopede	145
	5.12. Bicorn	147
	5.13. Piriform	149
	5.14. Devil's Curve	151
	5.15. Folia	151
	5.16. Cassinian Oval	153
	5.17. Cartesian Oval	155
	5.18. Dürer's Conchoid	157

		Page
6.	Algebraic Curves of High Degree	160
	6.1. Epitrochoid	160
	6.2. Hypotrochoid	165
	6.3. Epicycloid	168
	6.4. Nephroid	170
	6.5. Hypocycloid	171
	6.6. Astroid	173
	6.7. Rhodonea	175
	6.8. Nephroid of Freeth	175
	6.9. Cayley's Sextic	178
	6.10. Bowditch Curve	178
7.	Transcendental Curves	184
	7.1. Sinusoidal Spiral	184
	7.2. Logarithmic Spiral	184
	7.3. Archimedean Spirals	186
	7.4. Euler's Spiral	190
	7.5. Involute of a Circle	190
	7.6. Epi Spiral	192
	7.7. Poinsot's Spirals	192
	7.8. Cochleoid	192
	7.9. Cycloid	192
	7.10. Quadratrix of Hippias	195
	7.11. Catenary	195
	7.12. Tractrix	199
References		201
Appendix A. Tables of Derived Curves		202
Appendix B. Further Reading		208
Index of Curve Names		216

ILLUSTRATIONS

		Page
1.	Illustration of Some Notation	xi
2.	Pedal Coordinates	3
3.	Polar Equation of a Line	6
4.	Angles	7
5.	Bipolar to Cartesian Coordinates	14
6.	Translation of Coordinates	16
7.	Curvature	23
8.	Center of Curvature	24
9.	Geometric Properties	37
10.	Involute	41
11.	Parallel Curves	42
12.	Inverse Curves	44
13.	Pedal Curves	46
14.	Negative-Positive Pedals	47
15.	Conchoid	49
16.	Strophoid	51
17.	Cissoid	53
18.	Isoptic	58
19.	Caustic	60
20.	Form-1 of the Parabola	68
21.	Parabolae	68
22.	Ellipse	74
23.	Hyperbola	74
24.	Semi-Cubical Parabola	87
25.	Tschirnhausen's Cubic	87
26.	Witch of Agnesi	93
27.	Cissoid of Diocles	93
28.	Pedals of a Parabola	97
29.	Right Strophoid	103
30.	Trisectrix of Maclaurin	103
31.	Folium of Descartes	109
32.	Trident of Newton	109

		Page
33.	Serpentine	112
34.	Limacon of Pascal	114
35.	Trisectrix	114
36.	Cardioid	120
37.	Lemniscate of Bernoulli	120
38.	Eight Curve	126
39.	Bullet Nose -1	126
40.	Bullet Nose -2	127
41.	Cross Curve	127
42.	Deltoid	135
43.	Conchoid of Nicomedes -1	135
44.	Conchoid of Nicomedes -2	136
45.	Kappa Curve	136
46.	Kampyle of Eudoxus	143
47.	Hippopede -1	143
48.	Hippopede -2	144
49.	Bicorn	148
50.	Piriform	148
51.	Devil's Curve	152
52.	Folium	152
53.	Cassinian Oval	154
54.	Cartesian Oval	156
55.	Dürer's Conchoid	158
56.	Epitrochoid -1	162
57.	Epitrochoid -2	163
58.	Epitrochoid -3	164
59.	Hypotrochoid -1	166
60.	Hypotrochoid -2	167
61.	Epicycloid	169
62.	Nephroid	169
63.	Hypocycloid	172
64.	Astroid	172

		Page
65.	Rhodonea -1	176
66.	Rhodonea -2	177
67.	Nephroid of Freeth	177
68.	Cayley's Sextic	180
69.	Bowditch Curve -1	181
70.	Bowditch Curve -2	182
71.	Bowditch Curve -3	183
72.	Logarithmic Spiral	185
73.	Archimedes' Spiral	187
74.	Fermat's Spiral	187
75.	Hyperbolic Spiral	188
76.	Lituus	188
77.	Archimedean Spiral	189
78.	Euler's Spiral	191
79.	Involute of a Circle	191
80.	Epi Spiral	193
81.	Poinsot's Spiral #1	194
82.	Poinsot's Spiral #2	194
83.	Cochleoid	196
84.	Curtate Cycloid	196
85.	Cycloid	197
86.	Prolate Cycloid	197
87.	Quadratrix of Hippias	198
88.	Catenary	200
89.	Tractrix	200

TABLES

		Page
1.	Representation of Curves	1
2.	Pedal Equations	4
3.	Bipolar Equations	5
4.	Intrinsic Equations	5
5.	(Generalized) Pedals of the Parabola	95
6.	Evolutes and Involutes	202
7.	Radials	202
8.	Inverses	203
9.	Pedals	204
10.	Strophoids	205
11.	Cissoids	205
12.	Roulettes	206
13.	Isoptics	206
14.	Orthoptics	207
15.	Catacaustics	207

ABSTRACT

This report is an illustrated study of plane algebraic and transcendental curves, emphasizing analytic equations and parameter studies. It is aimed at the undergraduate student who has mastered analytic geometry and calculus. There are seven chapters. The first chapter provides a quick resumé of analytic facts about curves. Chapter 2 discusses the general subject of derived curves. The succeeding five chapters discuss more than 60 special plane curves in varying detail. Each curve is illustrated. An extensive chronological bibliography is provided in an appendix.

INTRODUCTION

There are a great many methods of studying individual curves, and a report which intends to discuss a large number of different curves must necessarily choose to emphasize only a few aspects. Lockwood,[1] for example, primarily discusses how to construct various curves; Yates,[2] on the other hand, provides a reference volume, with the usual advantages and problems of this approach. Both books suffer from a lack of illustrative material.

Chapter 1 provides a quick resumé (with few derivations) of facts required to understand the rest of the volume, and Chapter 2 discusses the general subject of derived curves. Succeeding chapters describe individual curves, roughly in order of increasing complexity, emphasizing analytic equations associated with the curves and illustrated parameter studies.

For each base curve, ten derived curves (including the curve itself) may be discussed:

The base curve	Pedal
Evolute	Inverse
Involute	Conchoid
Radial	Strophoid
Parallel	Cissoid

As a prerequisite, the reader is assumed to be familiar with the differential and integral calculus.

The illustrations in this report were prepared on a Cal Comp digital incremental plotter.

This work was performed at Lawrence Livermore Laboratory, Livermore, California, and depended heavily on the computer facilities available there. I am most grateful to Dr. Frederick N. Fritsch of that Laboratory for his thoughtful review and critical commentary on all the draft material. I am indebted to Mrs. Camille Muir for her painstaking efforts in typing the manuscript.

NOTATION

x, y	=	rectangular coordinates.
		$U(x,y) = 0$ will represent an arbitrary rectangular equation.
r, θ	=	polar coordinates.
		$\rho(r,\theta) = 0$ will represent an arbitrary polar equation.
r_1, r_2	=	bipolar coordinates.
		$\gamma(r_1,r_2) = 0$ will represent an arbitrary bipolar equation.
r, p	=	pedal coordinates.
		$\pi(r,p) = 0$ will represent an arbitrary pedal equation.
s, φ	=	intrinsic coordinates (Whewell).
s, ρ	=	intrinsic coordinates (Cesáro).
t	=	parameter.
		$F(t) = (f(t), g(t))$ will represent an arbitrary parametric function.
φ	=	inclination of tangent.
ψ	=	angle between tangent and radius vector to the point of tangency.
s	=	arc length.
p	=	distance from origin to tangent.
r	=	radial distance.
R	=	radial ray.
L	=	length.
A	=	area.
O	=	origin.
V	=	volume.
Σ	=	surface area.
V_x	=	volume of revolution about x-axis.
Σ_x	=	surface area of revolution about x-axis.
ρ	=	radius of curvature.
K	=	curvature.
ν	=	angle between two lines.
d	=	distance.
m	=	slope.
O, P, Q	=	point.
α, β	=	coordinates of the center of curvature.

-x-

$$f_x = \frac{\partial f}{\partial x}.$$
$$f'(x) = \frac{df}{dx}.$$
$$\dot{f} = \frac{df}{dt}.$$

(a, b, c) [when referring to the coordinates of a point] is equivalent to t = a, x = b, and y = c.

Figure 1 illustrates some of this notation.

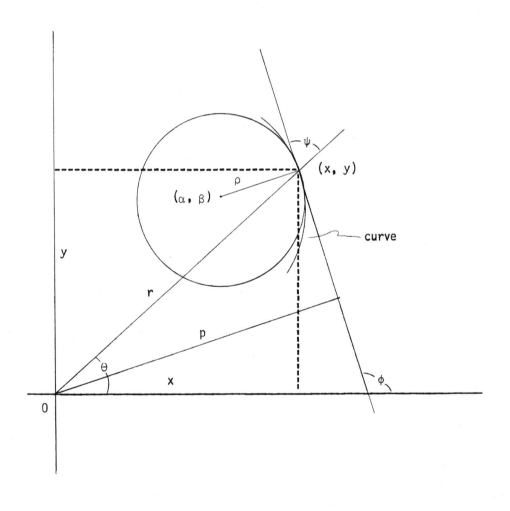

Figure 1. Illustration of Some Notation

A CATALOG OF
SPECIAL PLANE CURVES

CHAPTER 1

PROPERTIES OF CURVES

1.1. Coordinate Systems

Of the various systems for representing plane curves,* six will be discussed in this section. Each system has its own advantages and difficulties, as indicated in Table 1. Most of the succeeding chapters will use the parametric representation within a Cartesian coordinate system; direct Cartesian representation and polar representation will occur somewhat less frequently; and pedal, bipolar, and intrinsic systems will be of infrequent use.

The first three systems of representation mentioned in Table 1 are well known, and require only a few comments. Coordinate axes for the Cartesian and parametric systems are a pair of perpendicular lines, the *abscissa* (x-axis) and the *ordinate* (y-axis). Coordinate axes

TABLE 1. Representation of Curves

System	Advantages
Parametric	Separation of variables, curve tracing, derivatives
Cartesian	Slope, distance calculation, translation
Polar	Central point, angles, rotation
Pedal	Curvature
Bipolar	Two central points
Intrinsic	Invariants

**Curve* is not defined until section 1.6 (page 20). Intuitively, a curve is composed of a finite set of differentiable segments.

for the polar system consist of a point (the *pole*) and a ray from this point (the *axis*). In this work, the polar pole and axis will be assumed to fall on the Cartesian origin and positive x-axis, respectively.

In the parametric system, coordinates of a curve are expressed independently as functions of a single variable, t, such as

1.1.1) ...
$$\begin{cases} x = f(t) \\ y = g(t). \end{cases}$$

Hereafter, t, f, and g, with or without subscripts and/or superscripts, will be reserved for the parametric representation of some curve. The vector equation

1.1.2) ... $X = F(t) = \bigl(f(t), g(t)\bigr)$

will also be frequently used, as are vector representations of points

$$P = (x, y).$$

Parametric methods are most important for our purposes, and will be heavily relied on.

The Cartesian and polar coordinate systems are basically point concepts; given any point P, there is one and only one set of coordinates (x, y) or (r, θ) for P. Pedal coordinates, however, are basically dependent on curves, and a point P may have many different pedal coordinates (r, p), depending on the curve under consideration.

Let O be a fixed point (the pedal point, or pole) lying at the origin, and let C be a differentiable curve (i.e., its tangent exists). At a point P on C whose pedal coordinates are desired, construct the tangent line L to C. The *pedal coordinates* of P (with respect to C and O) are the radial distance r from O to P and the perpendicular distance p from 0 to L (Figure 2). For a different curve C_1 through P, r is, of course, the same, but p may very well be different. Furthermore, if C does not have a tangent at P (e.g., if P is an isolated point, or cusp), then the pedal coordinates of P do not exist. Some examples of pedal equations are given in Table 2.

Let O_1 and O_2 be two fixed points (the poles) a distance 2c apart. The line segment $L = O_1O_2$ is termed the base line, and the bisector of L is then known as the center. The *bipolar coordinates* of a point P are the distances r_1 and r_2 from O_1 and O_2, respectively. Now, O_1, O_2, and P form a triangle, so r_1, r_2, and c must satisfy the inequalities

$$1.1.3) \quad \begin{cases} r_1 + r_2 \geq 2c, \\ |r_1 - r_2| \leq 2c. \end{cases}$$

Further, since r_1, r_2, and c are all assumed to be positive, any equation in bipolar coordinates describes a locus that is symmetric about L; conversely, a locus that is not symmetric about some line cannot have a bipolar equation. Examples of bipolar equations are given in Table 3.

It is sometimes desirable to write the equation of a curve in such a manner as to be independent of certain coordinate transformations. The type of transformations being considered are those that

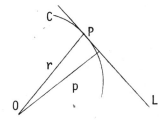

Figure 2.   Pedal Coordinates

preserve length and angle; all the transformations discussed below within and between coordinate systems fit this definition. Such a transformation will also preserve area, arc length, curvature, number of singularities, etc. Whewell introduced a system involving arc length s and tangential angle ϕ, while Cesáro gave a system involving arc length s and radius of curvature ρ. Since $\rho \, d\phi = ds$, by definition, it is evident that these are related. Examples of both are given in Table 4.

As an example of these systems of representation, consider the straight line, with Cartesian equation

1.1.4) ...　　　　$ax + by + c = 0, \quad (a^2 + b^2 \neq 0)$.

There are two frequently used parametric forms of this equation. Let $P_1 = (x_1, y_1)$ and $P_2 = (x_2, y_2)$ be two points. The two-point form of the linear equation may be written (in vector notation) as

1.1.5) ...　　　　$X = P_1 t + P_2 (1 - t)$,

TABLE 2. Pedal Equations

Curve	Pedal Point	Pedal Equation
Parabola	Focus	$p^2 = ar$
Circle	Center	$pa = r^2$
Ellipse	Focus	$\dfrac{b^2}{p^2} = \dfrac{2a}{r} - 1$
Lemniscate	Center	$pa^2 = r^3$
Astroid	Center	$r^2 + 3p^2 = a^2$

TABLE 3. Bipolar Equations

Curve	Polar Points	Bipolar Equation
Ellipse	Foci	$r_1 + r_2 = 2d, \; d \geq c$
Hyperbola	Foci	$r_1 - r_2 = \pm 2d$
Lemniscate	Foci	$r_1 r_2 = c^2$
Oval of Cassini	Foci	$r_1 r_2 = k^2$
Oval of Descartes	Foci	$mr_1 \pm nr_2 = k$

TABLE 4. Intrinsic Equations

Curve	Whewell Equation	Cesáro Equation
Astroid	$s = a \cos 2\phi$	$4s^2 + \rho^2 = 4a^2$
Cardioid	$s = a \cos \frac{1}{3} \phi$	$s^2 + 9\rho^2 = a^2$
Circle	$s = a\phi$	$\rho = a$
Catenary	$s = a \tan \phi$	$s^2 + a^2 = a\rho$
Cycloid	$s = a \sin \phi$	$s^2 + \rho^2 = a^2$
Tractrix	$s = a \ln \sec \phi$	$a^2 + \rho^2 = a^2 \exp(2s/a)$

while the one-point form is

1.1.6) ... $\quad X = P_1 + \Gamma t,$

with $\Gamma = (\lambda, \mu)$ being a parameter dependent on the slope of the line. This latter form may be connected with 1.1.4 by the relations

1.1.7) ...
$$\begin{cases} \lambda = -\dfrac{b}{\sqrt{a^2+b^2}} = -\dfrac{1}{\sqrt{1+m^2}} \\ \mu = \dfrac{a}{\sqrt{a^2+b^2}} = \dfrac{-m}{\sqrt{1+m^2}} \end{cases}.$$

In polar coordinates, the line has equation

1.1.8) ... $r \cos(\theta - \alpha) = p$,

where p is the distance from the origin to the line and α is the angle between the axis and this perpendicular (Figure 3). This may be related to 1.1.4 by

1.1.9) ... $a = \cos \alpha, \quad b = \sin \alpha, \quad c = -p,$

yielding

1.1.10) ... $x \cos \alpha + y \sin \alpha = p$.

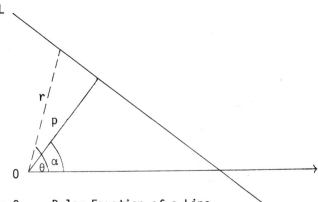

Figure 3.   Polar Equation of a Line

1.2 Angles

Three angles are of importance in coordinate geometry: the slope angle ϕ, the radial angle θ, and the tangential-radial angle ψ. Since the angle ν between lines is of frequent usage, it is also described here. Definitions are in terms of a point $P_0 = (x_0, y_0)$ on a line L; comparable definitions for a curve may be obtained using the tangent line (if it exists).

Let a Cartesian-coordinate system be given. The slope angle ϕ of a line L is defined to be the angle formed by L and the x-axis, taken clockwise from L (Figure 4); if L is parallel to the x-axis, then ϕ is taken to be zero. The radial angle θ is the angle between the radial line R between O and P_0 and the x-axis, taken clockwise from R. Finally, ψ is defined as the angle between R and L, taken clockwise from L. Note that all three angles are in the interval $[0, \pi)$.

Slope. The *slope* m of a line L is defined to be the tangent of angle ϕ. Using the standard equations 1.1.4, 1.1.6, and 1.1.8 for the line, m is found to be given by

1.2.1) ... $$m = \tan \phi = -\frac{a}{b} = \frac{\mu}{\lambda} = -\cot \alpha.$$

Two lines L_1 and L_2 (with appropriate equations), possessing slopes

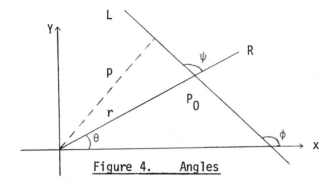

Figure 4.   Angles

m_1 and m_2 respectively, are perpendicular if any of the following relations be true:

$$1.2.2) \quad \begin{cases} m_1 m_2 + 1 = 0 \\ a_1 a_2 + b_1 b_2 = 0 \\ \mu_1 \mu_2 + \lambda_1 \lambda_2 = 0 \\ \cos(\phi_1 - \phi_2) = 0 \\ \cos(\alpha_1 - \alpha_2) = 0 \end{cases}$$

They are parallel if one of the next set of relations be true:

$$1.2.3) \quad \begin{cases} m_1 = m_2 \\ a_1 b_2 = a_2 b_1 \\ \mu_1 \lambda_2 = \mu_2 \lambda_1 \\ \phi_1 = \phi_2 \\ \alpha_1 = \alpha_2 \end{cases}$$

Tangent Lines. Anticipating the definition of a curve somewhat, suppose C is a curve differentiable at a point P_0 (i.e., the tangent line to C at P_0 exists), and let us find the equation of the tangent line. We will assume the curve to have one of the following equations: $U(x, y) = 0$, $F = (f(t), g(t))$, or $r = \rho(\theta)$.

Now, the slope of a curve, as is well known, is given by

$$m = \frac{dy}{dx}.$$

Differentiating the Cartesian form,

$$U_x dx + U_y dy = 0 ,$$

yields

1.2.4) ... $\quad \dfrac{dy}{dx} = -\dfrac{U_x}{U_y} ,$

where

$$U_x = \dfrac{\partial U}{\partial x} \quad \text{and} \quad U_y = \dfrac{\partial U}{\partial y} .$$

The parametric form presents no problems; it gives

1.2.5) ... $\quad \dfrac{dy}{dx} = \dfrac{dg(t)}{df(t)} = \dfrac{dg(t)}{dt} \Big/ \dfrac{df(t)}{dt} .$

The polar form, however, needs an auxiliary set of equations; namely,

$$x = r \cos \theta \quad y = r \sin \theta .$$

Using these (their derivation is quite simple), it is easy to discover that

1.2.6) ... $\quad m = \dfrac{\sin \theta \, dr + r \cos \theta \, d\theta}{\cos \theta \, dr - r \sin \theta \, d\theta} .$

Now, each of these results may be substituted into the appropriate equation for a line, to result in the equations for the tangent line to C at P_0. These equations are, respectively,

1.2.7) ... $\quad U_{x_0} x + U_{y_0} y - (U_{x_0} x_0 + U_{y_0} y_0) = 0 ;$

1.2.8) ... $\quad X = F(t_0) - t \dfrac{dF(t_0)}{\sqrt{df^2(t_0) + dg^2(t_0)}}$

$$(x,y) = \left(f_0 - \dfrac{t \, df_0}{\sqrt{df_0^2 + dg_0^2}} , \; g_0 - \dfrac{t \, dg_0}{\sqrt{df_0^2 + dg_0^2}} \right) ;$$

and

1.2.9) ... $\quad r\cos(\theta - \alpha) = r_0 \cos(\theta_0 - \alpha)$,

with

$$\cos \alpha = -m \quad \text{(from 1.2.6)}$$

and

$$P_0 = (x_0, y_0) = (f_0, g_0) = (f(t_0), g(t_0)) = (r_0, \theta_0) \; .$$

Radial Angle. The *radial angle*, θ, is, of course, one of the coordinate variables in the polar system. It may be connected to the Cartesian system by

1.2.10) ... $\quad \tan \theta = \frac{y}{x}$.

Tangential-Radial Angle. As may be seen from Figure 4, there must be a functional relationship between ϕ, θ, and the *tangential-radial angle* ψ. This is obviously

1.2.11) ... $\quad \psi + \theta = \phi$.

Taking tangents yields

$$\tan \psi = \tan(\phi - \theta) = \frac{\tan \phi - \tan \theta}{1 + \tan \phi \tan \theta} \; .$$

But $\tan \phi = m$ and m is given by 1.2.6 . Some algebra will now yield

1.2.12) ... $\quad \tan \psi = r \left(\dfrac{d\theta}{dr} \right)$,

or

1.2.13) ... $\quad \sin \psi = \dfrac{r}{\sqrt{r^2 + \left(\dfrac{dr}{d\theta}\right)^2}}$.

We shall return to the subject of ψ in a later section (1.7).

Angle Between Lines. If two lines, L_1 and L_2, have slopes m_1 and m_2, respectively, then the angle between them (clockwise from L_2) is given by

$$1.2.14) \quad \begin{cases} \tan \nu = \dfrac{m_2 - m_1}{1 + m_1 m_2} \;, \\[1em] \cos \nu = \dfrac{a_1 a_2 + b_1 b_2}{\sqrt{(a_1^2 + b_1^2)(a_2^2 + b_2^2)}} \;, \\[1em] \nu = \pi + \alpha_2 - \alpha_1 \;. \end{cases}$$

Angle Between Curves. If

$$x = f(t)$$
$$y = g(t)$$

represents a curve C, then the direction cosines of the tangent to C at a point P are given by

$$1.2.15) \quad \begin{cases} \cos \alpha = \pm \dfrac{\dot{f}}{\sqrt{\dot{f}^2 + \dot{g}^2}} \\[1em] \cos \beta = \pm \dfrac{\dot{g}}{\sqrt{\dot{f}^2 + \dot{g}^2}} \end{cases}$$

(where dots represent derivatives with respect to t). Hence, if

$$x = f_1(t_1) \qquad x = f_2(t_2)$$
$$y = g_1(t_1) \qquad y = g_2(t_2)$$

represent two curves C_1 and C_2, the angle $\delta(t_1, t_2)$ between the curves (that is, the angle between the tangent to C_1 for a given value of t_1 and the tangent to C_2 for a given value of t_2) is given by

$$1.2.16) \quad \cos \delta = \pm \dfrac{\dot{f}_1 \dot{f}_2 + \dot{g}_1 \dot{g}_2}{\sqrt{(\dot{f}_1^2 + \dot{g}_1^2)(\dot{f}_2^2 + \dot{g}_2^2)}} \;.$$

1.3 Changes Between Coordinate Systems

In this section, five transformation sets between four coordinate systems will be developed. These systems, and their coordinate variables, are

$$\begin{array}{ll} \text{Cartesian} & (x, y) \\ \text{Polar} & (x, \theta) \\ \text{Pedal} & (r, p) \\ \text{Bipolar} & (r_1, r_2) \end{array}$$

The transformations between Cartesian and polar coordinates are well known, and are given by

1.3.1) ... $\qquad x = r \cos \theta \quad , \qquad y = r \sin \theta$

and

1.3.2) ... $\qquad r = \sqrt{x^2 + y^2} \quad , \qquad \theta = \arctan(\frac{y}{x}) \quad .$

If $U(x, y) = 0$ is the Cartesian equation of a curve, then its polar equation is found, using 1.3.1, to be $U(r \cos \theta, r \sin \theta) = 0$; similar rules work the opposite way. Thus, the cardioid

$$r = 2a(1 + \cos \theta)$$

has Cartesian equation

$$\sqrt{x^2 + y^2} = 2a \left(1 + \frac{x}{\sqrt{x^2 + y^2}} \right) \quad ,$$

or

$$(x^2 + y^2 - 2ax)^2 = 4a^2(x^2 + y^2) \quad .$$

Suppose a curve C is given by $U(x, y) = 0$. Then, letting $U_x = \frac{\partial U}{\partial x}$, etc.,

$$U_x \, dx + U_y \, dy = 0 ,$$

or

$$\frac{dy}{dx} = - \frac{U_x}{U_y} .$$

Hence, the tangent line L through a point (x_0, y_0) on C is given by

$$U_{y_0}(y - y_0) + U_{x_0}(x - x_0) = 0 .$$

Calculating the distance from the origin to L yields

1.3.3) ... $$p = \frac{xU_x + yU_y}{\sqrt{U_x^2 + U_y^2}} ;$$

r is, of course, given by

1.3.4) ... $$r = \sqrt{x^2 + y^2} .$$

Eliminating x and y from these two equations, plus the defining relation $U(x,y) = 0$, yields the pedal equation for C (the exact equation depends on U, of course).

If C has equation $F(t) = (f(t), g(t))$, then the tangent is

$$(y - g)df = (x - f)dg ,$$

so

1.3.5) ... $$p = \frac{g \, df - f \, dg}{\sqrt{df^2 + dg^2}} .$$

In this case, t must be eliminated. Now suppose $r = r(\theta)$ is the equation for C. Then (Figure 4)

1.3.6) ... $$p = r \sin \psi = \frac{r^2}{\sqrt{r^2 + \left(\frac{dr}{d\theta}\right)^2}} \;;$$

here, ψ and θ must be eliminated, with the aid of 1.2.13.

Reversing these steps is not well defined in general, due to the invariance under rotation of the pedal system.

As an example, consider the ellipse with polar equation $ar(1 + e \cos \theta) = b^2$, with $b^2 = a^2(1 - e^2)$. Solving for r, differentiating, and eliminating θ yields

$$r = \frac{b^2}{a(1 + e \cos \theta)}$$

and

$$\frac{dr}{d\theta} = r \left(\frac{e \sin \theta}{1 + e \cos \theta}\right) = \frac{r}{b} \sqrt{2ar - r^2 - b^2} \;.$$

But, (1.3.6)

$$p = \frac{r^2}{\sqrt{r^2 + \left(\frac{dr}{d\theta}\right)^2}} \;,$$

so, after a little algebra, we have

$$p^2 = \frac{b^2 r}{2a - r} \;,$$

the desired equation.

It now remains to derive the transformations to and from bipolar coordinates and Cartesian and polar coordinates.

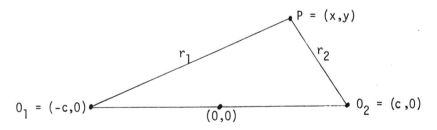

Figure 5. Bipolar to Cartesian Coordinates

The conversion from bipolar to Cartesian coordinates may be conveniently done with the origin either at the midpoint of $O_1 O_2$ (so $O_1 = (-c, 0)$ and $O_2 = (c, 0)$) or at one of the centers. In the first case (Figure 5)

1.3.7) ... $\quad r_1^2 = (x + c)^2 + y^2, \quad r_2^2 = (x - c)^2 + y^2.$

Subtracting gives

1.3.8) ... $\quad r_1^2 - r_2^2 = 4xc, \quad$ or $\quad x = \dfrac{r_1^2 - r_2^2}{4c}.$

Substitution of this value into 1.3.7, and solving for y leads to

1.3.9) ... $\quad y = \pm \dfrac{1}{4c} \left[16c^2 r_1^2 - (r_1^2 - r_2^2 + 4c^2) \right]^{1/2};$

the $\pm$ sign reflects the symmetry conditions about $O_1 O_2$.

From $r^2 = x^2 + y^2$ and $\tan \theta = y/x$, the transformation to polar coordinates can be easily derived, to yield

1.3.10) ... $\quad r = \dfrac{\sqrt{2}}{2} \left[r_1^2 + r_2^2 - 2c^2 \right]^{1/2},$

1.3.11) ... $\quad \tan \theta = \left[\dfrac{8c^2(r_1^2 + r_2^2 - 2c^2)}{r_1^2 - r_2^2} - 1 \right]^{1/2}.$

Furthermore, from 1.3.7, the reverse transformation may also be found:

1.3.12) ... $\quad r_1^2 = r^2 + c^2 + 2cr \cos \theta, \quad r_2^2 = r^2 + c^2 - 2cr \cos \theta.$

If $O_1 = (0,0)$ and $O_2 = (2c, 0)$, the same method leads to the equations

1.3.13) ... $\quad r_1^2 = x^2 + y^2, \quad r_2^2 = (x - 2c)^2 + y^2,$

1.3.14) ... $\begin{cases} x = \frac{1}{4c}(r_1^2 - r_2^2 + 4c^2) , \\ y^2 = r_1^2 - \frac{1}{16c^2}(r_1^2 - r_2^2 + 4c^2)^2 , \end{cases}$

1.3.15) ... $r = r_1, \quad \tan\theta = \left[\frac{16c^2 r_1^2}{r_1^2 - r_2^2 + 4c^2} - 1\right]^{1/2} .$

1.4. Changes Within Coordinate Systems

In this section, we shall consider, for the Cartesian and polar systems, the three methods of changing coordinates that preserve length and angle; these are the *translation*, the *rotation*, and the *reflection*.

Let $P = (x,y) = (r,\theta)$ be a point in the x-y (or r-θ) system with origin 0. If a change is made to an x'-y' (or r'-θ') system with center 0', what are the new coordinates of $P = (x',y') = (r',\theta')$?

First, consider a translation; that is, the new axes are parallel to the old ones, but the origin has moved to a point $0_0 = (x_0, y_0) = (r_0, \theta_0)$ with respect to the old system (Figure 6).

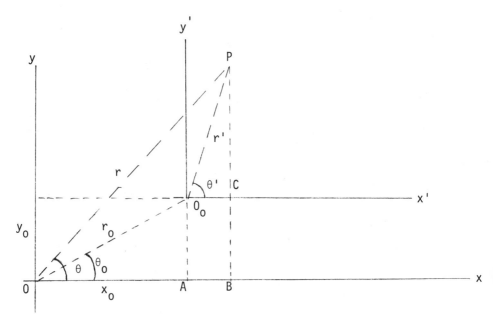

Figure 6. Translation of Coordinates

In the Cartesian case, it is clear that

1.4.1) ... $\quad x = x' + x_o, \quad\quad y = y' + y_o.$

The polar result is a little more complicated. Draw O_oA and PB perpendicular to the x-axis (and, thus, the x'-axis). Now

$$\angle PO_oO = \pi - \theta' + \theta_o.$$

Hence, by the law of cosines,

$$r^2 = r_o^2 + (r')^2 - 2r_o r' \cos(\pi - \theta' + \theta_o),$$

or

$$r^2 = (r_o)^2 + (r')^2 + 2r_o r' \cos(\theta_o - \theta').$$

Furthermore,

$$\tan\theta = \frac{r \sin\theta}{r \cos\theta} = \frac{PC + CB}{AB + OA} = \frac{r'\sin\theta' + r_o\sin\theta_o}{r'\cos\theta' + r_o\cos\theta_o}.$$

Hence, we have

1.4.2) ... $\begin{cases} r^2 = (r_o)^2 + (r')^2 + 2r_o r' \cos(\theta_o - \theta) \\[1em] \tan\theta = \dfrac{r'\sin\theta' + r_o\sin\theta_o}{r'\cos\theta' + r_o\cos\theta_o}. \end{cases}$

Note that neither of these is linear.

Rotations are especially straightforward in polar form. If a coordinate rotation through an angle θ_o is performed, we have

1.4.3) ... $\begin{cases} x = x'\cos\theta - y'\sin\theta, \\ y = x'\sin\theta + y'\cos\theta, \end{cases}$

and

1.4.4) ... $\quad r = r', \quad\quad \theta = \theta' + \theta_o.$

Finally, if e = ±1, a reflection may be obtained by

1.4.5) ... $x = ex'$, $y = -ey'$

or by

1.4.6) ... $r = er'$, $\theta = -e\theta'$.

1.5. Distances

For the Cartesian, parametric, and polar systems, we must now present distance formulae of the types listed below:

>point-point
>point-line
>point-curve
>line-line .

__Point-Point.__ By the Pythagorean formula, the distance d between points $P_1 = (x_1, y_1)$ and $P_2 = (x_2, y_2)$ is given by

1.5.1) ... $d^2 = (x_2 - x_1)^2 + (y_2 - y_1)^2$.

In polar coordinates, as may easily be verified using the law of cosines, this result is

1.5.2) ... $d^2 = r_1^2 + r_2^2 - 2r_1 r_2 \cos(\theta_1 - \theta_2)$.

One special case of interest is the radial distance r between a point P and the origin. This is, of course, given by

$$r^2 = x^2 + y^2 .$$

__Point-Line.__ Let L be the line $ax + by + c = 0$ (or $r\cos(\theta - \alpha) = P$) and P_o be a point. As is shown in elementary calculus courses, the (perpendicular) distance from P_o to L is given by

1.5.3) ... $$d = \frac{ax_0 + by_0 + c}{\sqrt{a^2 + b^2}} = r_0 \cos(\theta_0 - \alpha) - P .$$

A particular example is given by the tangential distance from the origin to the tangent line of a curve; this results in

1.5.4) ... $$P = \frac{xU_x + yU_y}{\sqrt{U_x^2 + U_y^2}} = \frac{y\, dx - x\, dy}{\sqrt{dx^2 + dy^2}} = \frac{r^2 d\theta}{\sqrt{dr^2 + r^2 d\theta^2}} .$$

Division of a Line Segment. To divide a line segment from P_1 to P_2 in the ratio $h_1:h_2$, $h_1 + h_2 \neq 0$, so that

$$\frac{P_1 P}{PP_2} = \frac{h_1}{h_2} ,$$

the coordinates of the point of division are given by

1.5.5) ... $$x = \frac{h_2 x_1 + h_1 x_2}{h_1 + h_2} , \qquad y = \frac{h_2 y_1 + h_1 y_2}{h_1 + h_2} .$$

Point-Curve. Let P_0 be a point, and suppose C is a curve defined by $F = (f(t), g(t))$. The distance from P_0 to C is given by

$$d^2 = (x_0 - f(t))^2 + (y_0 - g(t))^2 .$$

To find the shortest distance from P_0 to C, this must be minimized; under the usual differentiability conditions, this is equivalent to $\frac{d(d)}{dt} = 0$. But, if $\frac{d(d)}{dt} = 0$, then $\frac{d(d^2)}{dt} = 2d \frac{d(d)}{dt} = 0$. Hence,

$$0 = \frac{d(d^2)}{dt} = \frac{d}{dt}\left[(x_0 - f(t))^2 + (y_0 - g(t))^2\right]$$

$$= -2(x_0 - f(t))\frac{df(t)}{dt} - 2(y_0 - g(t))\frac{dg(t)}{dt} .$$

This is equivalent to solving

$$1.5.6) \quad 0 = (x_0 - f(t)) \frac{df(t)}{dt} + (y_0 - g(t)) \frac{dg(t)}{dt}$$

for t. A similar result for polar coordinates, if C is given by $r = r(\theta)$, may be found from 1.5.2 by differentiating

$$d^2 = r_0^2 + r^2(\theta) - 2r_0 r(\theta) \cos(\theta_0 - \theta) :$$

$$0 = 2d \frac{d(d)}{d\theta} = 2r \frac{dr}{d\theta} - 2r_0 \frac{dr}{d\theta} \cos(\theta_0 - \theta) - 2r_0 r \sin(\theta_0 - \theta) .$$

This last equation,

$$1.5.7) \quad (r - r_0 \cos(\theta_0 - \theta)) \, dr = r_0 r \sin(\theta_0 - \theta) \, d\theta$$

must now be solved for (r,θ).

Line-Line. If two lines are not parallel, the distance between them is, of course, zero. If they are parallel, pick any arbitrary point on one line and use 1.5.3 .

1.6. Curve

We must now define exactly what is meant by a curve; the development closely follows that of Kreyszig,[3] and is based on the parametric representation of a curve. We start with arcs, and put arcs together to build a curve.

Now, since differentiation is very important, the prime characteristic of an arc is just this. The second requirement excludes singular points.

A pair of functions $(f(t), g(t))$, t in some open interval I, form a legal representation of an arc if

1) The functions $f(t)$ and $g(t)$ are twice continuously differentiable.
2) For all $t \in I$, at least one of $\frac{df}{dt}$ and $\frac{dg}{dt}$ is nonzero.
3) For all points s and t in I, $(f(s), g(s)) = (f(t), g(t))$ if and only if $s = t$.

It should be clear that a given arc may have more than one legal representation. For example, the parabola may be represented by $y = 2at$, $x = at^2$ or by $y = 2a \tan \hat{t}$, $x = a \tan^2 \hat{t}$.

A pair of functions $(f(t), g(t))$, t in some interval I, is a legal representation of a curve C if there exists a finite set of points $t_0, t_1, \ldots, t_n$ in I such that (f,g) is a legal representation of an arc for $t_i < t < t_{i+1}$, for $i = 0, 1, \ldots, n-1$.

Note that this definition is a very limited one (a more general definition might replace "finite" by "countable"), but our primary goal is the study of particular curves, and this definition is thus quite satisfactory.

A curve is *continuous* if its defining equations are continuous. A curve is *algebraic* if its defining Cartesian equations is algebraic; otherwise it is *transcendental*.

1.7. Curvature

Arc Length. In Cartesian coordinates, the formula that is developed in elementary calculus for arc length is

1.7.1) ... $$\left(\frac{ds}{dx}\right)^2 = 1 + \left(\frac{dy}{dx}\right)^2$$

or

1.7.2) ... $$(ds)^2 = (dx)^2 + (dy)^2 \qquad \text{(parametric form)}.$$

Integrating gives

1.7.3) ... $$s = \int \left[1 + \left(\frac{dy}{dx}\right)^2\right]^{1/2} dx,$$

or

1.7.4) ... $$s = \int \left[(dx)^2 + (dy)^2\right]^{1/2} dt.$$

If $t = \theta$, this can be expressed in polar form as

1.7.5) ... $$(ds)^2 = (dr)^2 + r^2(d\theta)^2$$

so that

1.7.6) ... $$s = \int \left[\left(\frac{dr}{d\theta}\right)^2 + r^2\right]^{1/2} d\theta.$$

The form in pedal coordinates,

1.7.7) ... $$ds = \frac{r\, dr}{\sqrt{r^2 - p^2}},$$

will be developed below. (Equation 1.7.16)

<u>Curvature</u>. Curvature is defined intrinsically using arc length and inclination ϕ of the tangent line (to some base line, usually the x-axis), by

1.7.8) ... $$K = \frac{d\phi}{ds}.$$

The radius of curvature is the reciprocal of the absolute value of the curvature,

1.7.9) ... $$\rho = \frac{1}{|K|} = \left|\frac{ds}{d\phi}\right|.$$

Since
$$K = \frac{d\phi}{ds} = \frac{d\phi}{dx} \bigg/ \frac{ds}{dx},$$

using, where the curve has equation $y = u(x)$,

1.7.10) ... $$\frac{d\phi}{dx} = \frac{d}{dx}(\arctan u'(x)) = \frac{1}{1 + [u'(x)]^2} u''(x),$$

it is easy to see that

1.7.11) ... $$K = \frac{u''(x)}{\{1 + [u'(x)]^2\}^{3/2}}.$$

In parametric form,

1.7.12) ... $$K = \frac{\dot{f}(t)\ddot{g}(t) - \ddot{f}(t)\dot{g}(t)}{[(\dot{f})^2 + (\dot{g})^2]^{3/2}}.$$

If the curve is $U(x,y) = 0$, and U_{xx}, U_{xy}, U_{yy} are the second partial derivatives,

1.7.13) ... $$K = \frac{U_{xx} U_y^2 - 2U_{xy} U_x U_y + U_{yy} U_x^2}{(U_x^2 + U_y^2)^{3/2}} .$$

By using the parametric form for K, and the transformation

$$x = r \cos \theta, \quad y = r \sin \theta$$

to polar coordinates, a little algebra yields the polar form of curvature,

1.7.14) ... $$K = \frac{r^2 + 2(r')^2 - rr''}{[r^2 + (r')^2]^{3/2}} ,$$

where

$$r' = \frac{dr}{d\theta} \quad \text{and} \quad r'' = \frac{d^2r}{d\theta^2} .$$

The expression for curvature in pedal coordinates is very simple. As shown in the preceding section, $ds^2 = dr^2 + r^2 d\theta^2$; and from equation 1.2.12., $\tan \psi = r \frac{d\theta}{dr}$, so (see Figure 7)

$$\left(\frac{ds}{dr}\right)^2 = 1 + r^2 \left(\frac{d\theta}{dr}\right)^2 = 1 + \tan^2 \psi = \sec^2 \psi .$$

Hence,

$$\frac{dr}{ds} = \cos \psi = t/r . \quad \text{Also,} \quad \frac{r \, d\theta}{ds} = \sin \psi ,$$

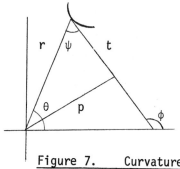
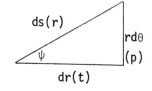

Figure 7.   Curvature

and $\cot \psi = \dfrac{dr}{r\, d\theta} = \dfrac{t}{p}$.

Combining these results gives

$$t = r\,\frac{dr}{ds} = \frac{p}{r}\,\frac{dr}{d\theta} ,$$

so that

$$\frac{d\theta}{ds} = \frac{p}{r^2} .$$

But $p = r \sin \psi$, so $dp = (\sin \psi)\, dr + r (\cos \psi)\, d\psi$, or $\dfrac{dp}{ds} = \dfrac{p}{r}\,\dfrac{dr}{ds} + t\,\dfrac{d\psi}{ds}$.
Therefore, rearranging and using $t\, ds = r\, dr$, we have

$$\frac{d\psi}{ds} = \frac{1}{r}\,\frac{dp}{dr} - \frac{p}{r^2} .$$

But, $K = \dfrac{d\phi}{ds} = \dfrac{d\psi}{ds} + \dfrac{d\theta}{ds}$, so we have, in the end,

1.7.15) ... $\quad K = \dfrac{1}{r}\,\dfrac{dp}{dr}$.

The radius of curvature, of course, is then given by $\rho = r\,\dfrac{dr}{dp}$.

As a byproduct, the promised expression for arc length can be easily given. From $ds^2 = dr^2 + r^2 d\theta^2$ and $r\, d\theta = \dfrac{p}{r}\, ds$, it is immediate that

1.7.16) ... $\quad ds^2 = \dfrac{r^2 dr^2}{r^2 - p^2}$.

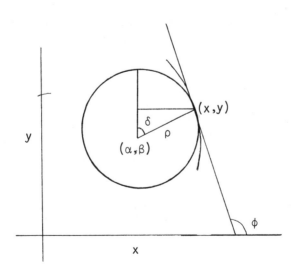

Figure 8. Center of Curvature

Center of Curvature. From Figure 8, it is clear that $\delta = \pi - \phi$, so

$$x - \alpha = \rho \sin \phi, \qquad y - \beta = -\rho \cos \phi,$$

and

$$\alpha = x - \rho \sin \phi, \qquad \beta = y + \rho \cos \phi.$$

But

$$\tan \phi = y', \text{ so } \sin \phi = \frac{y'}{\sqrt{1 + (y')^2}} \text{ and } \cos \phi = \frac{1}{\sqrt{1 + (y')^2}}.$$

Using this, with equation 1.7.11, results in

1.7.17) ... $\qquad \alpha = x - \frac{y'[1 + (y')^2]}{y''}, \qquad \beta = y + \frac{1 + (y')^2}{y''}.$

For parametric equations, $\tan \phi = \frac{dy}{dx} = \frac{\dot{g}(t)}{\dot{f}(t)}$

$$\sin \phi = \frac{\dot{g}}{\sqrt{(\dot{f})^2 + (\dot{g})^2}}, \quad \text{and} \quad \cos \phi = \frac{\dot{f}}{\sqrt{(\dot{f})^2 + (\dot{g})^2}},$$

where dots denote derivatives with respect to t.

Using

$$\rho = \frac{[(\dot{f})^2 + (\dot{g})^2]^{3/2}}{\dot{f}\ddot{g} - \ddot{f}\dot{g}}$$

results in

1.7.18) ... $\begin{cases} \alpha = f(t) - \dfrac{[(\dot{f})^2 + (\dot{g})^2]\dot{g}}{\dot{f}\ddot{g} - \ddot{f}\dot{g}}, \\ \\ \beta = g(t) + \dfrac{[(\dot{f})^2 + (\dot{g})^2]\dot{f}}{\dot{f}\ddot{g} - \ddot{f}\dot{g}}. \end{cases}$

The expression is quite messy in polar coordinates, but will be given anyway:

1.7.19) ...
$$\alpha = r\cos\theta - \frac{r^2 + (r')^2}{r^2 + 2(r')^2 - rr''} d(r\sin\theta)$$

$$\beta = r\sin\theta + \frac{r^2 + (r')^2}{r^2 + 2(r')^2 - rr''} d(r\cos\theta) .$$

1.8. Mensuration

In this section is given, without proof, a collection of measurement formulae.

<u>Area</u>. Suppose $y = u(x)$ is a curve, $a \leq x \leq b$. To find the area between the curve and the x-axis, it is necessary to integrate:

1.8.1) ...
$$A = \int_a^b u(x)dx .$$

$$= \int_{t_0}^{t_1} g(t)\dot{f}(t)dt .$$

Suppose $r = \rho(\theta)$, $\alpha \leq \theta \leq \beta$. To find the area of the arc,

1.8.2) ...
$$A = \int_\alpha^\beta \frac{1}{2}[\rho(\theta)]^2 d\theta .$$

<u>Volume of a Solid of Revolution</u>. Suppose $y = u(x)$, $a \leq x \leq b$, and suppose $y \geq 0$ on this interval. Then the volume of revolution about the x-axis is

1.8.3) ...
$$V_x = \pi\int_a^b [u(x)]^2 dx .$$

$$= \pi\int_{t_0}^{t_1} [g(t)]^2 \dot{f}(t) dt.$$

Surface Area of a Solid of Revolution. The function may be expressed as $y = u(x) \geq 0$, $a \leq x \leq b$, or as $x = f(t)$, $y = g(t) \geq 0$ for $a \leq t \leq b$. In either case, the area (about the x-axis) is

1.8.4) ... $$\Sigma_x = 2\pi \int_a^b y \, ds \ .$$

This breaks down as

$$\Sigma_x = 2\pi \int_a^b u(x) \sqrt{1 + [u'(x)]^2} \, dx$$

and as

$$\Sigma_x = 2\pi \int_a^b g(t) \sqrt{[\dot{f}(t)]^2 + [\dot{g}(t)]^2} \, dt$$

respectively.

About the y-axis, the formulae are

1.8.5) ... $$\Sigma_y = 2\pi \int_a^b x \, ds \ ,$$

$$\Sigma_y = 2\pi \int_a^b x \sqrt{1 + [u'(x)]^2} \, dx \ ,$$

and

$$\Sigma_y = 2\pi \int_a^b f(t) \sqrt{[\dot{f}(t)]^2 + [\dot{g}(t)]^2} \, dt$$

respectively.

<u>Volume</u>. Let $F(x,y) = z \geq 0$ in a region R. Then

$$V = \iint_R F(x,y) \, dA \ .$$

If R is defined as $a \leq x \leq b$, $c(x) \leq y \leq d(x)$, then

$$V = \int_a^b \int_{c(x)}^{d(x)} F(x,y) \, dy \, dx \ .$$

<u>Surface Area</u>. Let $F(x,y) = t \geq 0$ in R. Then

1.8.6) ... $$\Sigma = \iint_R \sqrt{1 + \left(\frac{\partial F}{\partial x}\right)^2 + \left(\frac{\partial F}{\partial y}\right)^2} \, dA \ .$$

1.9. Geometry

There are two different types of geometric quantities connected with curves. Those properties that depend only on the behavior of the curve in a small neighborhood of a point are termed *properties in the small*, and are generally expressed analytically by using derivatives of the equation of the curve at the point. Those properties that depend on the behavior of major portions of the curve are termed *properties in the large*, and are generally expressed by means of integrals (see section 1.8).

In this section, we will be discussing a curve C which can be represented by the Cartesian equation $U(x,y) = 0$ or the parametric equations $x = f(t)$, $y = g(t)$. No further mention will be made of this fact. Many of the properties discussed below are illustrated in Figure 9.

<u>Intercepts</u>. An *intercept* of C is a point $P = (x,y)$ on C that lie on the x-axis or y-axis (i.e., the product $xy = 0$). For example, the ellipse $9x^2 + 4y^2 = 36$ has the four intercepts $(0,3)$, $(-2,0)$, $(0,-3)$, $(2,0)$.

Extent. The *extent*, or *range* of C measures the upper and lower bounds to values that x and y may take on C. For example, the ellipse mentioned above has x-extent [2,2] and y-extent [-3,3]. However, the parabola $y^2 = 4x$ has x- and y-extents [0,∞) and (-∞,∞), respectively.

Branch. A *branch* of a curve is a set of connected arcs of C (in the sense of section 1.6). Frequently, a branch will be terminated by a singular point, point of discontinuity, or point at infinity at either or both ends. For example, consider the hyperbola xy = 1. Among the branches of the hyperbola are the following:

i) the branch for which $0 < x \leq 1$.
ii) the branch for which $1 \leq x < \infty$.
iii) the branch for which $0 < x < \infty$.

Note that the last includes both the first two.

Loop. A *loop* of C is an arc that completely encloses an area. That is, if the curve is represented by

$$x = f(t)$$
$$y = g(t) ,$$

there exist values t_0 and t_1 such that

i) C has an arc for $t_0 \leq t \leq t_1$;
ii) $f(t_0) = f(t_1)$ and $g(t_0) = g(t_1)$.

Note that, since it is demanded that a loop be an arc, it cannot cross itself (among other things it cannot do).

Critical Points. The *critical points* of C are those points at which the curve has a *maximum, minimum,* or *point of inflection*. The first two of these are dependent on the position of the coordinate axes; we may discuss maxima and minima in the y-direction or in the x-direction. Points of inflection are independent of the coordinate axes.

In the discussion below, $y' = dg(t) / df(t)$.

A *maximum value of y* occurs at a point $P = (x_0, y_0)$ on C if
i) $y' = 0$ or ∞ for $x = x_0$;
ii) there is a small neighborhood (a,b) of x_0 in which $y' > 0$ for $a < x < x_0$ and $y' < 0$ for $x_0 < x < b$.

If the function representing C is analytic in a neighborhood of P, a necessary and sufficient condition that P be a maximum is that
i) $y' = 0$ for $x = x_0$;
ii) the lowest order derivative that is not zero at P is of even order and negative.

The definition of a minimum value of y is quite similar to the above. $P = (x_0, y_0)$ is a *minimum value of y* if
i) $y' = 0$ or ∞ for $x = x_0$;
ii) there is a small neighborhood (a,b) of x_0 in which $y' < 0$ for $a < x < x_0$ and $y' > 0$ for $x_0 < x < b$.

If the function representing C is analytic in a neighborhood of P, a necessary and sufficient condition that P be a minimum is that
i) $y' = 0$ for $x = x_0$;
ii) the lowest order derivative that is not zero at P is of even order and positive.

As an example, consider the curve given by $y = x^n$, $n > 1$. Now, $y' = nx^{n-1}$, so $y' = 0$ at $x = 0$. Further, if n is even, $(0,0)$ is a minimum value of y (since $y' < 0$ for $x < 0$ and $y' > 0$ for $x > 0$). Now, the k-th derivative of y is given by $y^{(k)} = \frac{n!}{(n-k)!} x^r$. Hence, $y^{(k)} = 0$ for $x = 0$, $k < n$. Thus, the first nonzero derivative of y is $y^{(n)}$; it is of even order and positive.

Statements similar to the above may be made for maximum and minimum values of x by interchanging x and y throughout.

A point $P = (x_0, y_0)$ is a *point of inflection* of C if
i) $y'' = 0$ or ∞ for $x = x_0$;
ii) there is a small neighborhood (a,b) of x_0 in which the sign of y'' for $a < x < x_0$ is opposite to the sign of y'' for $x_0 < x < b$.

A point of inflection marks a change in sign of the curvature (1.7.11); the center of curvature "moves" from one side of C to the other.

A typical example is the curve given by $y = x^n$, $n > 1$ odd. Here, $y'' = n(n-1) x^{n-2}$; $y'' = 0$ at $x = 0$, $y'' < 0$ for $x < 0$, and $y'' > 0$ for $x > 0$.

Symmetry. Two points P and Q are *symmetric* with respect to a third point T if T bisects the line segment joining P and Q. P and Q are symmetric with respect to a line L if L is the perpendicular bisector of the line segment joining P and Q. Finally, a curve C is symmetric with respect to a point T (or a line L) if, for every point P on C, there is another point Q on C such that P and Q are symmetric with respect to T (or L).

In Cartesian coordinates C is symmetric with respect to the origin if its equation $U(x,y) = 0$ is unaltered when both variables x and y are replaced by their negatives. C is symmetric with respect to the x-axis (y-axis) if its equation is unaltered when $y(x)$ is replaced by its negative $-y(-x)$. Examples of these cases are the hyperbola $xy = 1$, and the parabolas $y^2 = x$ and $y = x^2$.

In polar coordinates, C [with equation $\rho(r,\theta) = 0$] is symmetric with respect to the origin (polar axis; the line $\theta = \pi/2$) if its equation remains unchanged when $r(\theta;\theta)$ is replaced by $-r(-\theta; \pi-\theta)$. Examples are the spiral $r^2 = \theta$ and the circles $r = \cos \theta$ and $r = \sin \theta$.

Asymptote. A line L is an *asymptote* to a curve C if L is a tangent to C at infinity. That is, a point P on C receding an infinite distance from the origin approaches indefinitely close to L. For example, the hyperbola $x^2 - y^2 = 1$ has as asymptotes the lines $x = y$ and $x = -y$.

A point O is an asymptote to a curve C if a point on C approaches arbitrarily close to O as it proceeds along C; an example is the spiral $r\theta = 1$, with the origin as asymptote.

Discontinuity. If a curve C is not continuous at a point P, P is (naturally) termed a *point of discontinuity* of C.

P is a *removable* point of discontinuity if the function can be made continuous by redefining it at P; both the right-hand and left-hand limits of C at P exist, and they are equal. An example is the curve $y = x \sin \frac{1}{x}$, which has a removable discontinuity at the origin.

P is an *ordinary* point of discontinuity if the limits from the right and the left exist but are unequal. An example of an ordinary nonremovable point of discontinuity occurs for $y = \arctan \frac{1}{x}$ at the origin; the right- and left-hand limits are $\pi/2$ and $-\pi/2$, respectively.

P is a *finite* point of discontinuity if the function that represents C is bounded in a neighborhood of P. An example of a finite non-ordinary non-removable discontinuity is the origin in $y = \sin \frac{1}{x}$, which approaches arbitrarily close to all points in $[-1, +1]$ of the y-axis in any neighborhood of the origin.

Finally, P is an *infinite* point of discontinuity if the function is unbounded near P. For example, $xy = 1$ at the origin.

Singularity. A point P on C is called an *ordinary* point if 1) y can be expressed as a continuous differentiable function of x at P, or 2) x can be expressed as a continuous differentiable function of y at P. Roughly, this means that C has a tangent at P that is a close approximation to C in a neighborhood of P; C does not cross itself at P; and P is not an isolated point. P is a *singular* point if it is not ordinary.

P is an *isolated* point (*hermit point; acnode*) of C if there is no other point of C in some neighborhood of P.

P is a *double* (*triple, quadruple, ...*) point of C if two (three, four, ...) arcs of C pass through P. P is a *multiple* point of C if more than one arc of C passes through P.

P is a *node* (*crunode*) of C if P is a double point of C and the two arcs that pass through P cross and have different tangents.

P is a *cusp* (*spinode*) of C if P is a double point of C whose two tangents coincide. P is a cusp of the first kind if there is an arc of C on each side of the double tangent in the neighborhood of P. P is a cusp of the second kind if the two arcs of C lie on the same side of the tangent in the neighborhood of P. P is a *double cusp* (*point of osculation; tacnode*) if the two arcs extend in both directions of the tangents.

If C can be represented in the neighborhood of P by a function $U(x,y) = 0$ with continuous second partial derivatives, then P is singular if $U_x = U_y = 0$ at P. Singular points may be classified by the sign of

1.9.1) $\quad\quad\quad \Delta = U_{xy}^2 - U_{xx}U_{yy}$

as follows (provided not all second order partial derivatives vanished):

$\quad\quad\quad \Delta < 0 \quad\quad\quad$ isolated point
$\quad\quad\quad \Delta = 0 \quad\quad\quad$ cusp
$\quad\quad\quad \Delta > 0 \quad\quad\quad$ node

If all the second order derivatives vanish, more complicated singularities may occur.

Envelope. The *envelope* of a one parameter family of curves is a curve that is tangent to every curve of the family. The equation of the family may be given in Cartesian coordinates by $U(x,y,c) = 0$; in a parametric form by $x = f(t,c)$ and $y = g(t,c)$; and so forth. To find the equation in the first of these conditions, it is necessary to eliminate the parameter c from the equation of the curve, $U(x,y,c) = 0$, and its partial derivative with respect to the parameter, $U_c(x,y,c) = 0$. Strictly speaking, this procedure is guaranteed only at points for which the sum $U_x^2 + U_y^2 \neq 0$; if both U_x and U_y are zero, the curve may have a singular point here.

In the parametric case, the parameter must be eliminated between the equations

1.9.2) ... $\begin{cases} x = f(t,c) \\ y = g(t,c) \\ 0 = f_t g_c - f_c g_t \end{cases}$

As an example, consider the family of ellipses

$$\frac{x^2}{c^2} + \frac{y^2}{(1-c)^2} = 1, \quad 0 \leq c \leq 1.$$

Taking the partial derivative and eliminating c results in

$$x^2 = c^3$$
$$y^2 = (1-c)^3 .$$

Hence, the envelope is the astroid

$$x^{2/3} + y^{2/3} = 1 .$$

In the parametric case, the equations are

$$x = c \cos t$$
$$y = (1-c) \sin t .$$

Finding partial derivatives and using equation 1.9.2 results in $c = \cos^2 t$. Substituting back gives the parametric equation of the astroid,

$$x = \cos^3 t$$
$$y = \sin^3 t .$$

Illustrations. The twelve curves given in Figure 9 illustrate the features discussed in this section. Here, we indicate what items each curve illustrates.

Figure 9a. The curve is the hyperbola $x^2 - y^2 = 1$. It has two x-intercepts, infinite extent, two arcs, and x-minimum and x-maximum points. It is symmetric with respect to both axes and the origin, is asymptotic to the lines $x = y$ and $x = -y$.

Figure 9b. The curve is $y^3 = x^2$. It has an intercept at the origin, infinite extent, two arcs, and no loops. There is a minimum point at the origin with $y' = \infty$. It is symmetric with respect to the y-axis, and there is a cusp of the first kind at that point.

Figure 9c. The curve is $y = x^3$. Again, the origin is the important point: intercept, point of inflection ($y'' = 0$), and point of symmetry. It has infinite extent and one arc.

Figure 9d. The curve is $y^3 = x^5$. This curve is much like the preceding one, except that the origin is a point of inflection with $y'' = \infty$.

Figure 9e. The curve $y = x \sin \frac{1}{x}$ illustrates a removable discontinuity at the origin. It has intercepts for $\frac{1}{x} = n\pi$, $n = \pm 1$, ± 2, ...; infinite extent, two arcs, and an infinite number of local maxima and minima in the interval $(-\frac{1}{\pi}, \frac{1}{\pi})$. The asymptote is $y = 1$.

Figure 9f. The spiral $r\theta = 1$ illustrates an asymptote at the origin.

Figure 9g. The family of ellipses $\frac{x^2}{a^2} + \frac{y^2}{(1-a)^2} = 1$ has the astroid $x^{2/3} + y^{2/3} = 1$ as an envelope. ($0 \leq a \leq 1$.)

Figure 9h. The curve is a pedal to the parabola $y^2 = 8x$, with pedal point $(-3,6)$; it has parametric equation

$$x = \frac{6t - 5t^2}{1 + t^2}$$

$$y = \frac{2t^3 - 3t + 6}{1 + t^2} .$$

There are three intercepts, approximately equal to $(0, 0, 6)$,*
$(1.2, 0, 1.869)$ $(-1.784, -8.36, 0)$. It has finite x-extent and
infinite y-extent. There is one asymptote: $x = -5$. This curve
also has a loop, with a node at $(-3, 6)$ for $t = \frac{1}{2}(3 \pm \sqrt{15})$.

Figure 9i. The curve is the semi-cubical parabola $27y^2 = 4x^3$. It primarily illustrates a cusp of the first kind at the origin ($dx/dy = \infty$ here).

Figure 9j. This parallel to $y = x^5$ has parametric equations

$$x = t \pm \frac{35\ t^4}{\sqrt{1 + 25t^8}}$$

$$y = t^5 \mp \frac{7}{\sqrt{1 + 25t^8}}.$$

It is symmetric with respect to the origin, and illustrates two cusps of the second kind and a node.

Figure 9k. This is a conchoid to the ellipse $9x^2 + 25y^2 = 225$, with $k = 3$ and pole $(0, 8)$. It has two loops, finite extent (in both x and y directions), and a double cusp at the origin.

Figure 9ℓ. The curve is a conchoid to the trisectrix of Maclaurin

$$x(x^2 + y^2) = 2(y^2 - 3x^2)$$

with $k = 3$ and pole $(3, 0)$. It has a triple point, which is a node for one pair of arcs and a double cusp for each of the other two pairs of arcs. The point $(3, 0)$ is a cusp of the first kind, twice.

---

*The notation (a, b, c) means $t = a$, $x = b$, and $y = c$.

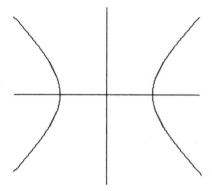
a. Hyperbola $x^2 - y^2 = 1$

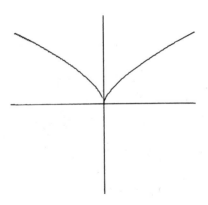
b. $y^3 = x^2$

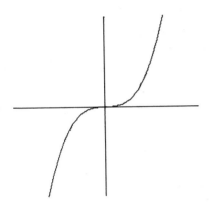
c. $y = x^3$

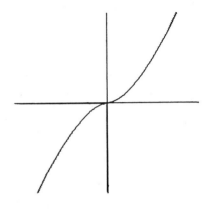
d. $y^3 = x^5$

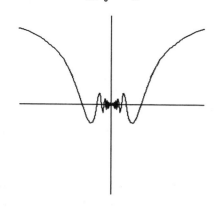
e. $y = x \sin \frac{1}{x}$

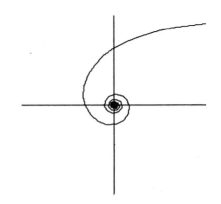
f. Spiral $r\theta = 1$

Figure 9. Geometric Properties

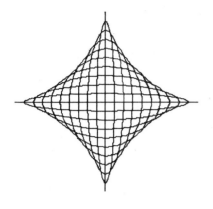

g. Envelope of Ellipses

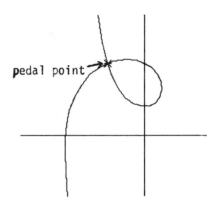

h. Pedal to Parabola

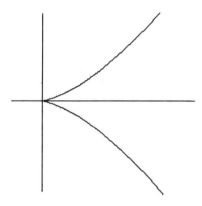

i. Semi-cubical Parabola

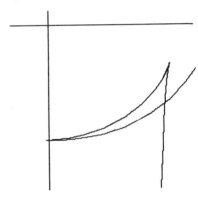

j. Parallel to $y = x^5$

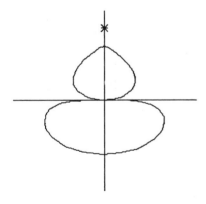

k. Conchoid to Ellipse

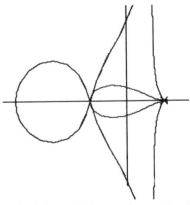

ℓ. Conchoid to Trisectrix of Maclaurin

Figure 9. (cont.) Geometric Properties

CHAPTER 2

TYPES OF DERIVED CURVES

Once a curve has been defined, it is possible to use some of its properties, together with auxiliary points, lines, and curves, to obtain new curves. This chapter discusses how this may be done, beginning with an arbitrary smooth curve defined in the various coordinate systems described above by

$$x = f(t)$$
$$y = g(t) ,$$

or

$$U(x,y) = 0 ,$$

or

$$r = \rho(\theta) ,$$

or

$$p = \pi(r) ,$$

or just as "C".

Auxiliary curves are described in the same manner, but subscripts are added; thus,

$$r = \rho_2(\theta)$$

or

$$x = f_1(t_1) .$$

No further mention of these facts will be deemed necessary; for example, when f is encountered in this chapter, it always means the function $x = f(t)$.

At various spots in this chapter, equations are written down with the implicit assumption that denominators are nonzero and all variables are finite. The justification for not discussing the contrary cases is the survey nature of this chapter. The interested reader should have no difficulty exploring these special cases, however.

2.1. Evolute, Involute, and Radial (Huygens, 1673; Tucker, 1864) *

Perhaps the most elementary derived curves are the evolute; its inverse, the involute; and the radial. The *evolute* of a curve, C, is the locus of its center of curvature. Parametric equations were derived earlier in 1.7.18:

$$2.1.1) \quad \begin{cases} x = f(t) - \dfrac{(\dot{f}^2 + \dot{g}^2)\,\dot{g}}{\dot{f}\ddot{g} - \ddot{f}\dot{g}} \\ \\ y = g(t) + \dfrac{(\dot{f}^2 + \dot{g}^2)\,\dot{f}}{\dot{f}\ddot{g} - \ddot{f}\dot{g}} \end{cases}.$$

Lines may be drawn from a fixed point, $O = (x_0, y_0)$ equal and parallel to the radii of curvature of C; the locus of end points is the *radial*. Clearly, the equation of the radial is the same as the equation of the evolute, translated to (x_0, y_0), so

$$2.1.2) \quad \begin{cases} x = x_0 - \dfrac{\dot{g}[(\dot{f})^2 + (\dot{g})^2]}{\dot{f}\ddot{g} - \ddot{f}\dot{g}} \\ \\ y = y_0 + \dfrac{\dot{f}[(\dot{f})^2 + (\dot{g})^2]}{\dot{f}\ddot{g} - \ddot{f}\dot{g}} \end{cases}$$

are the equations.

---

*Names in parentheses identify early investigators.

If a line, L, rolls (as a tangent), without slipping, along a fixed curve, C, any fixed point P on L is an *involute* of C. It can be shown that if S is the evolute of C, then C is an involute of S. All involutes of C are parallel (see 2.2).

Let C_0 be a point on C for which P lies on C, for $t = t_0$ (see Figure 10). The equation for the involute of C will be derived from this point.

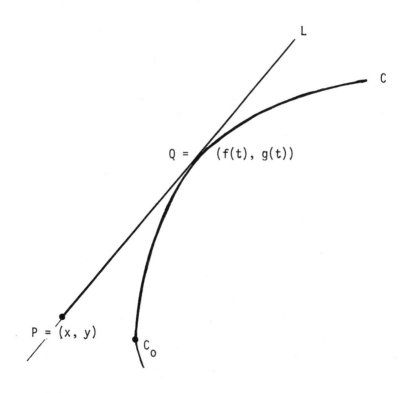

Figure 10. Involute

The equation of the line, L, in Figure 10 is

2.1.3) ... $\quad y - g(t) = \frac{dg}{df}(x - f(t))$.

Since the distance PQ = s, the length of arc C_0Q of C measured from C_0, we have

2.1.4) ... $\quad (x - f(t))^2 + (y - g(t))^2 = s^2$.

Solving 2.1.3 and 2.1.4 simultaneously yields the equation of the involute,

2.1.5) ... $\begin{cases} x = f(t) - \dfrac{s\, df}{\sqrt{df^2 + dg^2}} \\ \\ y = g(t) - \dfrac{s\, dg}{\sqrt{df^2 + dg^2}} \end{cases}$.

Notice that the negative square root must be used here.

2.2. Parallel Curves (Leibnitz, 1692)

If P is a variable point on C, the locus of points Q_1 and Q_2 that lie a distance ± k units from P along a line perpendicular to C define curves *parallel* to C. There are two branches.

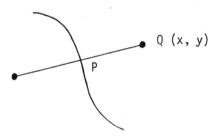

Figure 11. Parallel Curves

Let $P = (f(t), g(t))$ and $Q = (x, y)$. Since PQ is perpendicular to C, the equation of PQ (Figure 11) is

2.2.1) ... $\qquad dg(y - g(t)) = -df(x - f(t))$.

Since $|\overline{PQ}| = k$, it is possible to say that

2.2.2) ... $\qquad (x - f(t))^2 + (y - g(t))^2 = k^2$.

Substituting from 2.2.1 yields, after a little algebra, the parametric equations for the parallels

2.2.3) ... $$\begin{cases} x = f(t) \pm \dfrac{k\, dg}{\sqrt{df^2 + dg^2}} \\ y = g(t) \mp \dfrac{k\, df}{\sqrt{df^2 + dg^2}} \end{cases}.$$

2.3. Inversion (Steiner, 1824)

Let O be a fixed point (center of inversion). Suppose a line L is drawn through O intersecting C at P, and let Q be a point on L so that

2.3.1) ... $\qquad \overline{OP} \cdot \overline{OQ} = k$, a constant .

Then P and Q are inverse points, and the locus of Q is an *inverse* of C with respect to O. k may be negative, in which case P and Q lie on opposite sides of O.

If O is the pole, and C has polar equation $r = \rho(\theta)$, then the inverse of C has polar equation

2.3.2) ... $\quad r\,\rho(\theta) = k$

or

2.3.3) ... $\quad r = k/\rho(\theta)$.

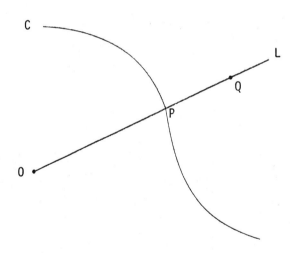

Figure 12. Inverse Curves

If $O = (x_0, y_0)$ and $P = (f(t), g(t))$, then L (Figure 12) has equation

2.3.4) ... $\quad y - y_0 = m(x - x_0)$

where

2.3.5) ... $\quad m = \dfrac{g(t) - y_0}{f(t) - x_0}$.

But

2.3.6) ... $\quad \overline{OP}^2 = (f(t) - x_0)^2 + (g(t) - y_0)^2 = (f(t) - x_0)^2(1 + m^2)$

and

2.3.7) ... $\quad \overline{OQ}^2 = (x - x_0)^2 + (y - y_0)^2 = (x - x_0)^2 (1 + m^2)$.

Combining these results yields

2.3.8) ... $\quad (x - x_0)(f(t) - x_0)(1 + m^2) = k$.

Solving this for x, and using 2.3.4 to find y, gives the equation for the inverse,

2.3.9) ... $\quad \begin{cases} x = x_0 + \dfrac{k(f(t) - x_0)}{(f(t) - x_0)^2 + (g(t) - y_0)^2} \\ \\ y = y_0 + \dfrac{k(g(t) - y_0)}{(f(t) - x_0)^2 + (g(t) - y_0)^2} \end{cases}$.

The circle with center O and radius k is known as the *circle of inversion*; points on this circle are invariant. Angles between two curves preserve magnitude, but reverse direction, on inversion. Asymptotes to a curve invert into tangents of the inverse.

Some curves invert into themselves; these are termed *anallagmatic* curves.

2.4. Pedal Curves (Maclaurin, 1718)

If C is a curve and O a point (the pedal point), the locus S of the foot of the perpendicular from O to a variable tangent to C is the *first (positive) pedal* of C with respect to O (Figure 13). The envelope of the line through a variable point P on C at right angles to OP is the *first negative pedal* of C with respect to O. Higher pedals of both varieties are defined by induction in the obvious manner.

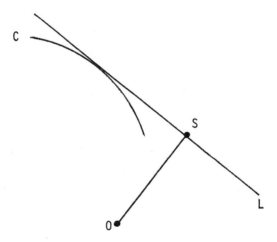

Figure 13. Pedal Curves

A similar definition in which "tangent" is replaced by "normal" is sometimes given to define the *normal pedal curve*.

It will now be shown that if S is the positive pedal of C, then C is the negative pedal of S.

From Figure 14 it is clear that p makes an angle of $(\phi - \pi/2)$ with the x-axis. Therefore,

2.4.1) ... $\tan \lambda = p \left(\dfrac{d\phi}{dp}\right)$

by the same reasoning leading to 1.2.12 .

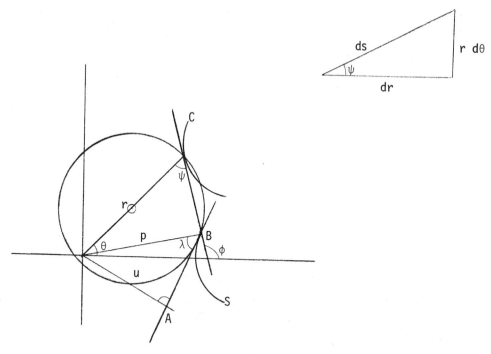

Figure 14. Negative-Positive Pedals

Therefore,

2.4.2) ... $$\tan \lambda \frac{dp}{ds} = p \frac{d\phi}{dp} \frac{dp}{ds} = p \frac{d\phi}{ds}$$

$$= p/\rho \text{ (since } \rho = ds/d\phi \text{ by definition)}$$

$$= p \frac{dp}{r\, dr} \quad \text{(by 1.7.15)}$$

$$= r \sin \psi \frac{dp}{r\, dr} = \sin \psi \frac{dp}{dr}.$$

Hence,

2.4.3) ... $$\tan \lambda = \sin \psi \frac{ds}{dr} = \tan \psi,$$

and so

2.4.4) ... $$\lambda = \psi.$$

Finally, this leads to

2.4.5) ... $$\frac{p}{r} = \frac{u}{p},$$

or

2.4.6) ... $$p^2 = ru.$$

Hence, the tangent to the pedal S is also tangent to the circle with r as diameter, and so C is the negative pedal of S.

Parametric Equation. The tangent line L to C has equation, as usual

2.4.7) ... $$y - g(t) = \frac{dg}{df}(x - f(t));$$

hence the perpendicular to L through $O(x_0, y_0)$ is

2.4.8) ... $$y - y_0 = -\frac{df}{dg}(x - x_0).$$

Solving these simultaneously leads (after some algebra) to the desired result:

$$2.4.9) \quad \begin{cases} x = \dfrac{x_0 df^2 + f(t)dg^2 + (y_0 - g(t))df\, dg}{df^2 + dg^2} \\[2ex] y = \dfrac{g(t)df^2 + y_0 dg^2 + (x_0 - f(t))df\, dg}{df^2 + dg^2} \end{cases}$$

<u>Pedal Equation</u>. If C has equation $r = \pi(p)$, then (2.4.6)

$$p^2 = ru = u\,\pi(p).$$

Thus, the pedal equation of the pedal to C is

$$2.4.10) \quad r^2 = p\,\pi(r).$$

2.5. Conchoid (Nicomedes, ca. 200 B.C.)

Let O be a fixed point, and let L be a line through O intersecting C at a point Q. The locus of points P_1 and P_2 on L such that

$$2.5.1) \quad P_1 Q = Q P_2 = k, \quad \text{a constant},$$

is a *conchoid* of C with respect to O (Figure 15).

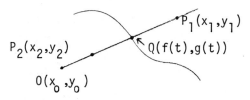

Figure 15. Conchoid

Parametric Equation. Once t is given, L has equation

2.5.2) ... $\quad y - y_0 = m(x - x_0)$

where

2.5.3) ... $\quad m = \dfrac{g(t) - y_0}{f(t) - x_0}$.

If $P(x,y)$ represents either P_1 or P_2; then, since P is on L,

2.5.4) ... $\quad y - g(t) = m(x - f(t))$.

However, since $(PQ)^2 = k^2$,

2.5.5) ... $\quad (x - f(t))^2 + (y - g(t))^2 = k^2$,

or, using 2.5.4 ,

2.5.6) ... $\quad (x - f(t))^2 = \dfrac{k^2}{1 + m^2}$.

Therefore,

2.5.7) ... $\quad x = f(t) \pm \dfrac{k}{\sqrt{1 + m^2}}$.

Using 2.5.3 and 2.5.4 results in the parametric equations of the conchoid,

2.5.8) ... $\begin{cases} x = f(t) \pm \dfrac{k(f(t) - x_0)}{\sqrt{(f(t) - x_0)^2 + (g(t) - y_0)^2}} \\ \\ y = g(t) \pm \dfrac{k(g(t) - y_0)}{\sqrt{(f(t) - x_0)^2 + (g(t) - y_0)^2}} \end{cases}$.

Polar Equation. If $O = (0,0)$ is the origin, then L has polar equation

$$\theta = \text{constant} .$$

Hence

2.5.9) ... $|P_1Q| = |\rho(\theta) - k|$,

so the polar equation for the conchoid is

2.5.10) ... $r = \rho(\theta) \pm k$.

2.6. Strophoid (Torricelli, 1645)

The strophoid involves one curve C and two points, O (the pole) and A (the fixed point). The locus of points P_1 and P_2 on a line L through O and intersecting C at a point Q such that

2.6.1) ... $P_2Q = QP_1 = QA$

is the *strophoid* of C with respect to O and A (Figure 16).

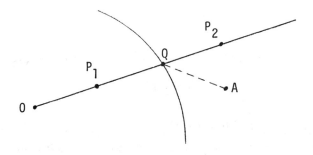

Figure 16. Strophoid

Parametric Equation. Let $O = (x_0, y_0)$ and $A = (x_1, y_1)$; C defined as usual. Now, L has equation

$$2.6.2) \quad \ldots \quad y - g(t) = m(x - f(t)) ,$$

where

$$2.6.3) \quad \ldots \quad m = \frac{g(t) - y_0}{f(t) - x_0} .$$

But

$$2.6.4) \quad \ldots \quad \overline{QA}^2 = (x_1 - f(t))^2 + (y_1 - g(t))^2 ,$$

and (where P represents either P_1 or P_2)

$$2.6.5) \quad \ldots \quad \overline{PQ}^2 = (x - f(t))^2 + (y - g(t))^2 .$$

Equating 2.6.4 and 2.6.5, using 2.6.2 and some algebra, results in

$$2.6.6) \quad \ldots \quad \begin{cases} x = f(t) \pm \dfrac{1}{\sqrt{1+m^2}} \left[(x_1 - f(t))^2 + (y_1 - g(t))^2\right]^{1/2} \\ \\ y = g(t) \pm \dfrac{m}{\sqrt{1+m^2}} \left[(x_1 - f(t))^2 + (y_1 - g(t))^2\right]^{1/2} \end{cases}$$

Polar Equation. If $O = (0,0)$, $A = (r_0, \theta_0)$, and C is represented by $r = \rho(\theta)$, then clearly the strophoid has equation

$$2.6.7) \quad \ldots \quad r = \rho(\theta) \pm \overline{QA} ,$$

or

2.6.8) ... $$r = \rho(\theta) \pm [r_0^2 + \rho^2(\theta) - 2r_0\rho(\theta)\cos(\theta - \theta_0)]^{1/2}.$$

2.7. Cissoid (Diocles, ca. 200 B.C.)

The cissoid involves two curves, C_1 and C_2, and a fixed point 0. Let Q_1 and Q_2 be the intersections of a line L through 0 with C_1 and C_2, respectively. The locus of points P on such lines such that

2.7.1) ... $$OP = OQ_2 - OQ_1 = Q_2Q_1$$

is the *cissoid* of C_1 and C_2 with respect to 0 (Figure 17).

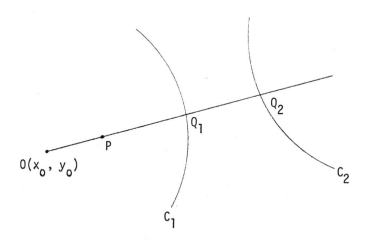

Figure 17. Cissoid

Parametric Equation. Choose t_1 as the basic parameter (i.e., the cissoid will be described parametrically in terms of parameter t_1). Then L has parametric equation (in terms of parameter v)

$$2.7.2) \quad \begin{cases} x = v[x_0 - f_1(t_1)] + f_1(t_1) \\ \\ y = v[y_0 - g_1(t_1)] + g_1(t_1) . \end{cases}$$

To find the intersection of L with C_2, these equations must be set equal to the parametric equations for C_2,

$$2.7.3) \quad \begin{cases} x = f_2(t_2) \\ \\ y = g_2(t_2) \end{cases}$$

and solved:

$$2.7.4) \quad \begin{cases} v[x_0 - f_1(t_1)] + f_1(t_1) = f_2(t_2) \\ \\ v[y_0 - g_1(t_1)] + g_1(t_1) = g_2(t_2) . \end{cases}$$

Letting

$$2.7.5) \quad \Delta x = f_1(t_1) - x_0 \quad \text{and} \quad \Delta y = g_1(t_1) - y_0 ,$$

v may be eliminated from 2.7.4 to obtain

$$2.7.6) \quad g_2(t_2)\Delta x - f_2(t_2)\Delta y = g_1(t_1)\Delta x - f_1(t_1)\Delta y .$$

Suppose this equation has a solution

2.7.7) ... $\quad t_2 = \sigma(t_1)$,

so that Q_2 has coordinates

2.7.8) ... $\quad \begin{cases} x = f_2(\sigma(t_1)) \\ \\ y = g_2(\sigma(t_1)) \end{cases}$

From this, the parametric equations of P are easily found to be

2.7.9) ... $\quad \begin{cases} x = x_0 + f_2(\sigma(t_1)) - f_1(t_1) \\ \\ y = y_0 + g_2(\sigma(t_1)) - g_1(t_1) \end{cases}$

<u>Polar Equation</u>. If $0 = (0,0)$, the polar equation of the cissoid is clearly

2.7.10) ... $\quad r = \rho_2(\theta) - \rho_1(\theta)$

<u>Cissoid to a Line</u>. An interesting special case occurs when C_2 is a line.

Suppose C_2 is $ax + by + c = 0$. If L goes through 0 and $C = C_1$, its equation is

2.7.11) ... $\quad (y - y_0)(f(t) - x_0) = (x - x_0)(g(t) - y_0)$;

solving simultaneously with C_2 results in the equation for the point Q_2 (provided L is not parallel to C_2):

$$2.7.12) \quad \begin{cases} x_2 = \dfrac{-c\Delta x + x_0 b\Delta y - y_0 b\Delta x}{a\Delta x + b\Delta y}, & a\Delta x + b\Delta y \neq 0 \\ \\ y_2 = \dfrac{-c\Delta y + y_0 a\Delta x - x_0 a\Delta y}{a\Delta x + b\Delta y}, \end{cases}$$

where

$$\Delta x = f(t) - x_0 \quad \text{and} \quad \Delta y = g(t) - y_0 .$$

Use of 2.7.9 finally results in the complete solution of this case:

$$2.7.13) \quad \begin{cases} x = \dfrac{b\Delta y(x_0 - \Delta x) - \Delta x(a\Delta x + by_0 + c)}{a\Delta x + b\Delta y} \\ \\ y = \dfrac{a\Delta x(y_0 - \Delta y) - \Delta y(b\Delta y + ax_0 + c)}{a\Delta x + b\Delta y} \end{cases}$$

provided

$$a\Delta x + b\Delta y \neq 0 .$$

2.8. Roulette (Besant, 1869)

If a curve C_1 rolls, without slipping, along another fixed curve C_2, any fixed point P attached to C_1 describes a *roulette*.

The term is also sometimes applied to the envelope of a fixed line attached to C_1, and to the locus of a variable point (such as the center of curvature of C_1 at the point of contact of the curves); however, these concepts will not be considered here.

Pedal Equation. Suppose C_2 is the x-axis, and C has equation $p = \pi(r)$ with respect to P. If N is the foot of the perpendicular from P to the x-axis, then

$$2.8.1) \quad \ldots \qquad p = N = y = r \frac{dx}{ds} \; ,$$

so the Cartesian equation of the roulette is

$$2.8.2) \quad \ldots \qquad y = \pi\!\left(y \frac{ds}{dx}\right) .$$

Theorem of Steiner. The following theorem of Steiner (quoted without proof) connects the areas and lengths of roulettes and pedal curves.[2]

Theorem: Let a point, P, rigidly attached to a closed curve, C, rolling on a line generate a roulette through one revolution of the curve. The area between the roulette and the line is twice the area of the pedal of C with respect to P; and the arc length of the roulette is equal to the corresponding arc length of the pedal.

A curve similar to the roulette is the *glissette* (Bernat, 1869), which is defined to be the locus of a point carried by a curve C as it slides between two given curves C_1 and C_2, or slides tangent to a given curve C_1 at a point. It can be shown that any glissette may also be defined as a roulette, so no more will be mentioned hereafter.

2.9. Isoptic (la Hire, 1704)

The locus of intersection of tangents to a curve C meeting at a constant angle α is an *isoptic* of C; if $\alpha = \pi/2$, the isoptic is termed an *orthoptic*.

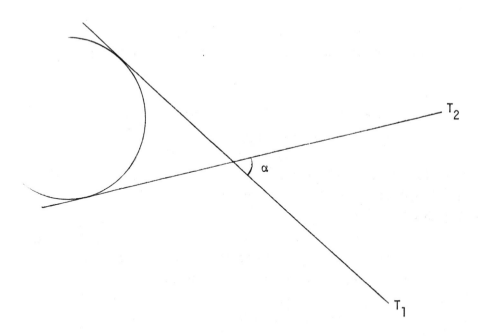

Figure 18. Isoptic

A tangent to C (Figure 18) has equation

2.9.1) ... $\quad y - g(t) = \frac{dg}{df}(x - f(t))$.

Since

$$\tan \alpha = \frac{m_2 - m_1}{1 + m_1 m_2}, \qquad (1.2.14) ,$$

the slope of T_2 may be found by solving for m_2; this yields

2.9.2) ... $\quad m_2 = \dfrac{m_1 + \tan \alpha}{1 - m_1 \tan \alpha} = \dfrac{dg + df \tan \alpha}{df - dg \tan \alpha}$.

It is now necessary to find points on C whose tangents have slope m_2, by solving

2.9.3) ... $\quad \dfrac{dg(t_o)}{df(t_o)} = m_2$

for t_o; T_2 then has equation

2.9.4) ... $\quad y - g(t_o) = m_2(x - f(t_o))$.

This may be solved simultaneously with 2.9.1 to arrive at the parametric equations of the isoptic,

2.9.5) ... $\begin{cases} x = \dfrac{m_2 f(t_o) - m_1 f(t) - g(t_o) + g(t)}{m_2 - m_1} \\[2ex] y = g(t) + m_1(x - f(t)) . \end{cases}$

2.10. Caustic (Tschirnhausen, Huygens, 1680)

A *caustic* of a given curve C is the envelope of light rays emitted from a point source S after reflection (*catacaustic*) or refraction (*diacaustic*) at C. If S is at infinity, the incident rays are parallel.

From Figure 19, $\overline{S}$ is the reflection of S in the tangent at T; the locus of $\overline{S}$ is called an *orthotomic* curve (secondary caustic). But QT is normal to the orthotomic, since $\overline{S}$ is the instantaneous center of motion of C, so the caustic is the evolute of the orthotomic.

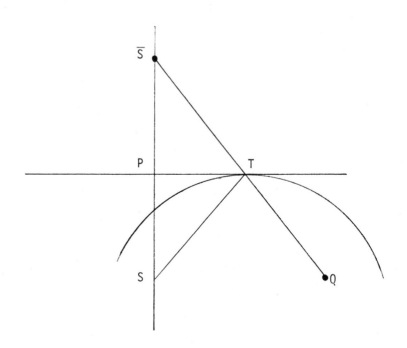

Figure 19. Caustic

CHAPTER 3

CONICS AND POLYNOMIALS

With this chapter, we begin a detailed analysis of individual curves. The chapters are ordered roughly by the degree of complexity of the curves. Each section describing an individual curve is divided into three parts, the first of which presents definitions of the curve, derivations of representative equations and illustrations for various values of any parameters present. The second part gives various geometric properties, and the last part similarly describes some analytic properties. No derivations are given in these two parts of each section. The properties described are listed below.

Geometric	Analytic		
Intercepts	x	$\dot{x}$	$\ddot{x}$
Local Extrema	y	$\dot{y}$	$\ddot{y}$
Points of Inflection	y''		
Range (Extent)	r		
Poles	θ		
Discontinuities	m		
Isolated Points	ψ		
Symmetries	p		
Asymptotes	$\dot{s}$ (or s)		
Loops	ρ		
Nodes	(α,β)		
Cusps	L,A (for loops only)		
	Σ_x, V_x		

3.1. Conics

A *conic* is defined geometrically as the locus of a point moving so that its distance from a fixed point (the *focus*) is proportional to its distance from a fixed line (the *directrix*); the curve is a parabola, an ellipse, or a hyperbola if the constant of proportionality (the *eccentricity*, e) is = 1, < 1, > 1 respectively.*

If the directrix is chosen as the line $x = 0$, and the focus as $F = (k, 0)$, then

$$e = \frac{FP}{DP}$$

where $P = (x,y)$ is on the conic and $D = (0,y)$ is on the directrix. But

$$FP = \sqrt{(x-k)^2 + y^2}$$

and

$$DP = |x| \, .$$

Hence

$$e = \frac{\sqrt{(x-k)^2 + y^2}}{|x|} ,$$

or

3.1.1) ... $\quad (1 - e^2) x^2 - 2kx + y^2 + k^2 = 0 \, .$

---

*For good general discussions of conics, see Yates[2] and Eisenhart.[4]

Thus, the conic has a second-degree algebraic equation; conversely, it can be shown that any second-degree algebraic equation represents a conic (including degenerate forms). Let the second degree equation be written as

3.1.2) ... $$ax^2 + by^2 + 2hxy + 2fx + 2gy + c = 0 ;$$

this equation has invariants (under translation and rotation) $a + b$, $\alpha = ab - h^2$, and determinant

$$D = \begin{vmatrix} a & h & f \\ h & b & g \\ f & g & c \end{vmatrix} .$$

Using α and D, a complete classification of second order curves may be obtained:

$\alpha = 0$	$D \neq 0$			parabola
	$D = 0$	$b \neq 0$	$g^2 - bc > 0$	two parallel real lines
			$g^2 - bc = 0$	two parallel coincident lines
			$g^2 - bc < 0$	two parallel imaginary lines
		$b=h=0$	$f^2 - ac > 0$	two parallel real lines
			$f^2 - ac = 0$	two parallel coincident lines
			$f^2 - ac < 0$	two parallel imaginary lines
$\alpha > 0$	$D = 0$			point ellipse
	$D \neq 0$		$-bD > 0$	real ellipse
			$-bD < 0$	imaginary ellipse
$\alpha < 0$	$D \neq 0$			hyperbola
	$D = 0$			two intersecting lines

In order to obtain a convenient equation for the conic, 3.1.1 is generally subjected to a change of coordinates. If $e = 1$, this transformation is

$$x' = x - \frac{1}{2}k$$

$$y' = y$$

to obtain the parabola

3.1.3) ... $\quad y^2 = 4ax, \quad$ where $\quad a = \frac{1}{2}k$,

with focus $F = (a,0)$ and directrix $(x + a = 0)$.
If $e \neq 1$, the translation is

$$x' = x - \frac{k}{1 - e^2}$$

$$y' = y$$

to obtain

3.1.4) ... $\quad \dfrac{x^2}{a^2} + \dfrac{y^2}{b^2} = 1$

where

$$a = \frac{ke}{1 - e^2} \quad \text{and} \quad b^2 = a^2(1 - e^2).$$

This *central conic* has focus $F = (-ae, 0)$ and directrix $(x + \frac{a}{e} = 0)$; by symmetry, $(ae, 0)$ and $(x - \frac{a}{e} = 0)$ are also a focus and directrix.

The discussion above does not include the circle, since e, by definition, is nonzero. However, the circle may be considered a limiting case of the ellipse, when $a = b$.

3.2. Circle

The *circle* is usually defined as the locus of a point P such that the distance from P to a fixed point O is constant. The point O is the *center*; the distance is the *radius*.

The circle is also a special case of the ellipse, for which $a = b$. It has Cartesian equation, if $O = (0,0)$,

3.2.1) ... $$x^2 + y^2 = a^2.$$

If $O = (h,k)$, the Cartesian equation is

$$(x - h)^2 + (y - k)^2 = a^2$$

or

$$x^2 + y^2 - 2hx - 2ky = a^2 - h^2 - k^2.$$

The parametric equations corresponding to 3.2.1 are

3.2.2) ... $$\begin{cases} x = a \cos t \\ y = a \sin t \end{cases} \quad -\pi \leq t \leq \pi.$$

The polar equation is

3.2.3) ... $r = a$,

and the pedal equation is $pa = r^2$.

Implicit equations are known for the circle; they are

$$s = a\phi \quad \text{(Whewell)},$$

$$\rho = a \quad \text{(Cesáro)}.$$

Geometry of the Circle.

Intercepts	$(0, a, 0);^*$ $(\pm\pi/2, 0, \pm a)$; $(\pm\pi, -a, 0)$
Extrema	$(0, \pm a, 0)$; $(\pm\pi/2, 0, \pm a)$
Extent	$-\pi \le t \le \pi$; $-a \le x \le a$; $-a \le y \le a$
Symmetries	$x = 0$; $y = 0$; $(0,0)$
Loops	$t \in [-\pi, \pi]$

Analysis of the Circle.

x	=	$a \cos t$	m	=	$-\cot t$
y	=	$a \sin t$	ψ	=	$\pi/2$
$\dot{x}$	=	$-a \sin t$	p	=	$-a$
$\ddot{x}$	=	$-a \cos t$	s	=	at
$\dot{y}$	=	$a \cos t$	ρ	=	$-a$
$\ddot{y}$	=	$-a \sin t$	(α,β)	=	$(0,0)$
y''	=	$-\dfrac{\csc^3 t}{a}$	L	=	$2a\pi$
			A	=	πa^2
r	=	a			
θ	=	t	V_x	=	$4\pi a^3/3$
			Σ_x	=	$4\pi a^2$

---

*Recall that this notation refers to values of (t, x, y).

3.3. Parabola

A *parabola* is the locus of a point P whose distance from a fixed point F (the *focus*) is equal to its distance from a fixed line $\mathcal{L}$ (the *directrix*). The intersection V of the parabola and the perpendicular from F to $\mathcal{L}$ (the axis) is the *vertex*. There is one parameter, a, equal to the distance between F and V. (See Figures 20 and 21.)

When studying special curves, it is generally convenient to choose the coordinate system in such a way as to minimize the complexity of the equation. In this manner, we may investigate the properties of the curve with less probability of getting lost in the algebraic manipulations. This was evident, for example, in the case of the circle, where we chose the origin to be the center of the circle.

In the case of the parabola, we choose the x-axis to coincide with the axis, and the origin to be at the vertex or at the focus. These will be referred to below as Form-1 and Form-2, respectively.

From Figure 20, it is evident that

$$d(P,D) = x + a = \sqrt{(x-a)^2 + y^2} = d(P,F) .$$

This, however, quickly reduces to the Cartesian equation for Form-1:

3.3.1) ... $$y^2 = 4ax .$$

The equation for Form-2 is immediate, by the translation $x' = x + a$:

3.3.2) ... $$y^2 = 4a(x + a) .$$

The parametric equation for Form-1 may be derived by setting $t = 2 \cot \theta$. Then,

$$t = 2 \cot \theta = \frac{2x}{y} = \frac{2}{y}\left(\frac{y^2}{4a}\right) = \frac{y}{2a} ,$$

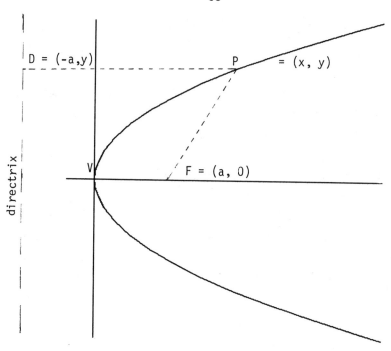

Figure 20. Form-1 of the Parabola

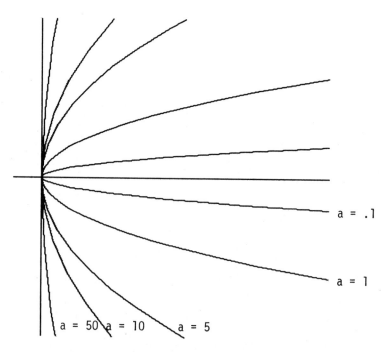

Figure 21. Parabolae
a = .1, 1, 5, 10, 50

or

$$y = 2at.$$

Now,

$$x = \frac{y^2}{4a} = \frac{4a^2t^2}{4a} = at^2,$$

so the equation is

3.3.3) ... $\begin{cases} x = at^2 & -\infty < t < \infty \\ \\ y = 2at. \end{cases}$

From this equation, Form-2 is immediate, by a translation:

3.3.4) ... $\begin{cases} x = a(t^2 - 1) & -\infty < t < \infty \\ \\ y = 2at. \end{cases}$

Now, let us transform equations 3.3.1 and 3.3.2 into polar coordinates. Form-1 proceeds from

$$y^2 = 4ax$$

$$r^2 = x^2 + y^2 = 4ax + x^2 = r^2 \cos^2\theta + 4ar\cos\theta.$$

Then

$$r(1 - \cos^2 \theta) = 4a \cos \theta$$

or

3.3.5) ... $\quad r \sin^2 \theta = 4a \cos \theta$.

It is also possible to convert Form-2 to polar form:

$$y^2 = 4a(x + a)$$

$$r^2 = x^2 + y^2 = 4ax + 4a^2 + x^2 = (x + 2a)^2 = (r \cos\theta + 2a)^2$$

or

3.3.6) ... $\quad r(1 - \cos \theta) = 2a$.

We may also derive a pedal equation from Form-2. Let

$$y = 2at \qquad\qquad x = a(t^2 - 1) .$$

Differentiating gives

$$y' = 2a \qquad\qquad x' = 2at .$$

Now,

$$p = \frac{(2at)^2 - 2a^2(t^2 - 1)}{\sqrt{(2at)^2 + (2a)^2}} = a\sqrt{1 + t^2}$$

and

$$r = \sqrt{a^2(t^2-1)^2 + (2at)^2} = a(t^2+1).$$

Hence,

3.3.7) ... $ar = p^2.$

Geometry of the Parabola.

	Form-1	Form-2
Intercepts	(0, 0, 0)	(0, -a, 0)
		(±1, 0, ±2a)
Extrema	(0, 0, 0)	(0, -a, 0)
Extent	$-\infty < t < \infty$	$-\infty < t < \infty$
	$0 \leq x < \infty$	$-a \leq x < \infty$
	$-\infty < y < \infty$	$-\infty < y < \infty$
Symmetries	y = 0	y = 0

Analysis of the Parabola.

Form-1	Form-2
$x = at^2$	$x = a(t^2 - 1)$
$y = 2at$	$y = 2at$
$\dot{x} = 2at$	$\dot{x} = 2at$
$\ddot{x} = 2a$	$\ddot{x} = 2a$
$\dot{y} = 2a$	$\dot{y} = 2a$

Analysis of the Parabola.

Form-1	Form-2
$\ddot{y} = 0$	$\ddot{y} = 0$
$y'' = -\dfrac{1}{2at^3}$	$y'' = -\dfrac{1}{2at^3}$
$r = at\sqrt{t^2 + 4}$	$r = a(1 + t^2)$
$\cot\theta = \dfrac{1}{2}t$	$\cot\theta = \dfrac{t^2 - 1}{2t}$
$m = \dfrac{1}{t}$	$m = \dfrac{1}{t}$
$\tan\psi = -\dfrac{t}{t^2 + 2}$	$\tan\psi = -\dfrac{1}{t}$
$p = \dfrac{at^2}{\sqrt{1+t^2}}$	$p = a\sqrt{1+t^2}$
$s = a\left[t\sqrt{1+t^2} + \ln(t+\sqrt{1+t^2})\right]$	$s = a\left[t\sqrt{1+t^2} + \ln(t+\sqrt{1+t^2})\right]$
$\rho = -2a(1+t^2)^{3/2}$	$\rho = -2a(1+t^2)^{3/2}$
$(\alpha,\beta) = (2a + 3at^2, -2at^3)$	$(\alpha,\beta) = (a + 3at^2, -2at^3)$

3.4. Ellipse

The *ellipse* was defined in section 3.1 as a conic with eccentricity e such that $0 < e < 1$. It can also be defined as the locus of a point P such that the sum of the distances from P to two fixed points (the *foci*) is constant. There are two parameters a and b, connected by

$$b^2 = a^2(1 - e^2).$$

The *center* of the ellipse is the midpoint of the line segment joining the foci. As with the parabola, there are two standard types. We choose the x-axis to coincide with the line joining the foci, and choose the origin to be at the center (Form-1) or a focus (Form-2). Some examples of ellipses are given in Figure 22.

Let the foci of the ellipse be $(\pm ae, 0)$, and let $P = (x,y)$ be a point on the ellipse. Then

$$\sqrt{(x - ae)^2 + y^2} + \sqrt{(x + ae)^2 + y^2} = 2a.$$

Squaring and collecting terms leads to the Cartesian equation for Form-1

3.4.1) ... $\quad \dfrac{x^2}{a^2} + \dfrac{y^2}{b^2} = 1.$

The Cartesian equation for Form-2 may be found by a translation $x' = x + ae$, and is

3.4.2) ... $\quad \dfrac{(x + ae)^2}{a^2} + \dfrac{y^2}{b^2} = 1.$

The parametric forms may be derived by means of the definition

$$\tan t = \frac{a}{b} \tan \theta.$$

But $\tan \theta = \dfrac{y}{x}$, and $y^2 = (a^2 - x^2)\dfrac{b^2}{a^2}$, so

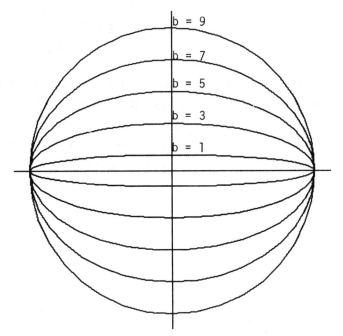

Figure 22. Ellipses
a = 9
b = 1, 3, 5, 7, 9

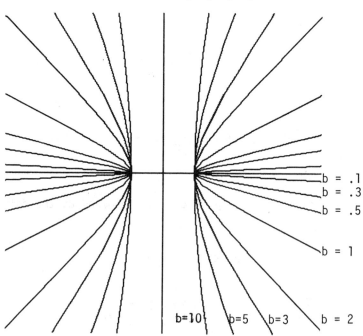

Figure 23. Hyperbolae
a = 2
b = .1, .3, .5, 1, 2, 3, 5, 10

$$\tan t = \sqrt{\frac{a^2 - x^2}{x^2}}.$$

But, $\tan^2 t = \sec^2 t - 1$, so

$$\cos t = \frac{x}{a}.$$

Hence, the parametric equations for Form-1 and Form-2 are

3.4.3) ... $\begin{cases} x = a \cos t \qquad -\pi \le t \le \pi \\ \\ y = b \sin t \end{cases}$

and

3.4.4) ... $\begin{cases} x = a(\cos t - e) \qquad -\pi \le t \le \pi. \\ \\ y = b \sin t \end{cases}$

Now, let us derive the polar equations of the ellipse. From 3.4.1, Form-1 is immediate:

3.4.5) ... $r^2 (a^2 \sin^2\theta + b^2 \cos^2\theta) = a^2 b^2$

or

$$r^2 (1 - e^2 \cos^2\theta) = b^2.$$

From 3.4.2, expanding and simplifying yields

$$a^2x^2 + a^2y^2 = a^2b^2 - a^2e^2b^2 - 2aeb^2x + x^2(a^2 - b^2)$$

$$a^2(x^2 + y^2) = a^2b^2(1 - e^2) - 2aeb^2x + a^2e^2x^2$$

$$a^2(x^2 + y^2) = (b^2 - aex)^2$$

$$ar = b^2 - aex = b^2 - aer\cos\theta$$

or

3.4.6) ... $ar(1 + e\cos\theta) = b^2$.

This last equation may be easily translated into pedal form:

$$r = \frac{b^2}{a(1 + e\cos\theta)}$$

$$\frac{dr}{d\theta} = r\left(\frac{e\sin\theta}{1 + e\cos\theta}\right) = \frac{r}{b}\sqrt{2ar - r^2 - b^2}.$$

But

$$p = \frac{r^2}{\sqrt{(r')^2 + r^2}}, \quad \text{so} \quad p^2 = \frac{b^2 r}{2a - r}.$$

Rearranging,

3.4.7) ... $\dfrac{b^2}{p^2} = \dfrac{2a}{r} - 1$.

Geometry of the Ellipse.

	Form-1	Form-2
Intercepts	$(0, a, 0)$	$(0, a-ae, 0)$
	$(\pi, -a, 0)$	$(\pi, -a-ae, 0)$
	$(\pm\pi/2, 0, \pm b)$	$(\pm \cos^{-1} e, 0, \pm b\sqrt{1-e^2})$
Extrema	$(0, a, 0)$	$(0, a-ae, 0)$
	$(\pi, -a, 0)$	$(\pi, -a-ae, 0)$
	$(\pm\pi/2, 0, \pm b)$	$(\pm\pi/2, \pm a-ae, \pm b)$
Symmetries	$x = 0;\ y = 0;\ (0,0)$	$y = 0$
Loops	$t \in [-\pi, \pi]$	$t \in [-\pi, \pi]$

Analysis of the Ellipse.

	Form-1		Form-2
x	$= a \cos t$	x	$= a(\cos t - e)$
y	$= b \sin t$	y	$= b \sin t$
$\dot{x}$	$= -a \sin t$	$\dot{x}$	$= -a \sin t$
$\ddot{x}$	$= -a \cos t$	$\ddot{x}$	$= -a \cos t$
$\dot{y}$	$= b \cos t$	$\dot{y}$	$= b \cos t$
$\ddot{y}$	$= -b \sin t$	$\ddot{y}$	$= -b \sin t$
y''	$= -\dfrac{b \csc^3 t}{a^2}$	y''	$= -\dfrac{b \csc^3 t}{a^2}$
r	$= \sqrt{a^2 \cos^2 t + b^2 \sin^2 t}$	r	$= a(1 - e \cos t)$
$\tan \theta$	$= \dfrac{b}{a} \tan t$	$\tan \theta$	$= \dfrac{b \sin t}{a(\cos t - e)}$

Form-1

$$m = -\frac{b}{a} \cot t$$

$$\tan \psi = -\frac{b}{ae \sin t \cos t}$$

$$p = -\frac{b}{\sqrt{1 - e^2 \cos^2 t}}$$

$$\dot{s} = a\sqrt{1 - e^2 \cos^2 t}$$

$$\rho = \frac{a^2 [1 - e^2 \cos^2 t]^{3/2}}{b}$$

$$(\alpha, \beta) = a e^2 \left(\cos^3 t, -\frac{a}{b} \sin^3 t\right)$$

$$L \approx \pi \left[c(a+b) - \sqrt{(a+3b)(b+3a)}\right]$$

$$A = \pi ab$$

$$V_x = \frac{4}{3} \pi ab^2$$

$$\Sigma_x = \pi b \left[2b + \frac{a^2}{\sqrt{b^2 - a^2}} \ln \frac{b + \sqrt{b^2 - a^2}}{b - \sqrt{b^2 - a^2}}\right]$$

Form-2

$$m = -\frac{b}{a} \cot t$$

$$\tan \psi = -\frac{b}{e \sin t}$$

$$p = -b \sqrt{\frac{1 - e \cos t}{1 + e \cos t}}$$

$$\dot{s} = a\sqrt{1 - e^2 \cos^2 t}$$

$$\rho = \frac{a^2 [1 - e^2 \cos^2 t]^{3/2}}{b}$$

$$(\alpha, \beta) = ae \left(e \cos^3 t - 1, -\frac{a}{b} \sin^3 t\right)$$

$$L \approx \pi \left[3(a+b) - \sqrt{(a+3b)(b+3a)}\right]$$

$$A = \pi ab$$

$$V_x = \frac{4}{3} \pi ab^2$$

$$\Sigma_x = \text{same as Form-1}$$

3.5. Hyperbola

The *hyperbola* was defined in section 3.1 as a conic with eccentricity e > 1. It can also be defined as the locus of a point P such that the difference of the distances of P from two fixed points (the *foci*) is constant. There are two parameters a and b, connected by

$$b^2 = a^2 (e^2 - 1) .$$

The hyperbola is termed *equilateral* if a = b.

The *center* of the hyperbola is the midpoint of the line segment joining the foci. There are three standard types, two of which are chosen exactly as the forms for the ellipse are chosen (section 3.5). Form-3 is chosen with the line joining the foci coinciding with the line x = y, and with the center coinciding with the origin. Form-3 is very common in applications. Some examples of hyperbolae are given in Figure 23.

The equations for Forms-1 and 2 are derived much as the analogous forms of the ellipse. The cartesian equations of the hyperbola are

3.5.1) ... $$\frac{x^2}{a^2} - \frac{y^2}{b^2} = 1 \qquad \text{(Form-1)}$$

and

3.5.2) ... $$\frac{(x + ae)^2}{a^2} - \frac{y^2}{b^2} = 1 \qquad \text{(Form-2)}.$$

The equation for Form-3 may be derived from 3.5.1, with a = b, by a rotation through π/4 radians. It is

3.5.3) ... $$xy = c^2$$

with

$$c = \frac{\sqrt{2}}{2} a.$$

The foci for these three forms are

(-ae, 0)	(+ae, 0)	(Form-1)
(-2ae, 0)	(0, 0)	(Form-2)
(ce, ce)	(-ce, -ce)	(Form-3)

The parametric equations are

3.5.4) ...
$$\begin{cases} x = a \sec t \\ y = b \tan t \end{cases} \quad -\pi \le t \le \pi \quad \text{(Form-1)}$$

3.5.5) ...
$$\begin{cases} x = a(\sec t - e) \\ y = b \tan t \end{cases} \quad -\pi \le t \le \pi \quad \text{(Form-2)}$$

3.5.6) ...
$$\begin{cases} x = ct \\ y = c/t. \end{cases} \quad -\infty < t < \infty \quad \text{(Form-3)}$$

The polar equations are

3.5.7) ... $\quad r^2 (e^2 \cos^2 \theta - 1) = b^2 \quad$ (Form-1)

3.5.8) ... $\quad ar(1 + e \cos \theta) = b^2 \quad$ (Form-2)

3.5.9) ... $\quad 2r^2 \sin 2\theta = a^2 e^2$. $\hfill$ (Form-3)

Pedal equations are

3.5.10) ... $\quad \dfrac{b^2}{p^2} = \dfrac{2a}{r} + 1$ $\hfill$ (Form-2)

3.5.11) ... $\quad p^2 (4r^4 + 15a^4 e^4) = 16 a^4 e^4$. $\hfill$ (Form-3)

For the hyperbola, geometric and analytic data will be given only for forms one and three.

Geometry of the Hyperbola.

	Form-1	Form-3
Intercepts	$(0, a, 0); (\pi, a, 0)$	---
Extrema	$(0, a, 0); (\pi, a, 0)$	---
Extent	$-\pi < t < \pi$	$-\infty < t < \infty$
	$-\infty < x < \infty$	$-\infty < x < \infty$
	$-\infty < y < \infty$	$-\infty < y < \infty$
Poles	$t = \pm \pi/2, \pm \pi$	$t = 0$
Discontinuity	$t = \pm \pi/2$	$t = 0$
Symmetries	$x = 0; y = 0; (0, 0)$	$(0, 0); x = \pm y$
Asymptotes	$bx = \pm ay$	$x = 0; y = 0$

Analysis of the Hyperbola.

Form-1

$$x = a \sec t$$
$$y = b \tan t$$
$$\dot{x} = a \sin t \sec^2 t$$
$$\ddot{x} = a \sec^3 t (1 + \sin^2 t)$$
$$\dot{y} = b \sec^2 t$$
$$\ddot{y} = 2b \sin t \sec^3 t$$
$$y'' = -\frac{b}{a} \csc^3 t$$
$$r = \sec t \sqrt{a^2 + b^2 \sin^2 t}$$
$$\tan \theta = \frac{b}{a} \sin t$$
$$m = \frac{b}{a} \csc t$$
$$\tan \psi = \frac{b}{ae^2} \cos^2 t \csc t$$
$$p = \frac{-ab \cos t}{\sqrt{a^2 \sin^2 t + b^2}}$$
$$\dot{s} = \sec t \sqrt{a^2 \sin^2 t + b^2}$$
$$\rho = \frac{-\sec^3 t [a^2 \sin^2 t + b^2]^{3/2}}{ab}$$
$$(\alpha, \beta) = ae^2 (\sec^3 t, -\frac{a}{b} \tan^3 t)$$

*From $t = 1$, for $t > 0$

Form-3

$$x = ct$$
$$y = c/t$$
$$\dot{x} = c$$
$$\ddot{x} = 0$$
$$\dot{y} = -\frac{c}{t^2}$$
$$\ddot{y} = \frac{2c}{t^3}$$
$$y'' = \frac{2}{ct^3}$$
$$r = \frac{c}{t} \sqrt{1 + t^4}$$
$$\tan \theta = \frac{1}{t^2}$$
$$m = -\frac{1}{t^2}$$
$$\tan \psi = -\frac{2t^2}{t^4 - 1}$$
$$p = \frac{2ct}{\sqrt{1 + t^4}}$$
$$s^* = c\sqrt{t^2+1} - c \ln\left[\frac{1 + \sqrt{1+t^2}}{t}\right]$$
$$\quad - c[\sqrt{2} - \ln(1 + \sqrt{2})]$$
$$\rho = \frac{c(1 + t^4)^{3/2}}{2t^3}$$
$$(\alpha, \beta) = \frac{c}{2t} \left(\frac{3t^4 + 1}{t^2}, t^4 + 3\right)$$

3.6. Power Function

We shall now consider the curve

3.6.1) ... $y = x^n$, $n \neq 0$.

No other form of this equation will be considered. There are several interesting special cases:

    i) n an integer.
    ii) $n = p/q$ rational.
    iii) n irrational.

The first of these is an important special case of the polynomial (see section 3.7). Illustrations can be found in Figures 9b, 9c, 9d, and 21.

Geometry of the Power Function.

	n even integer or n rational p even, q odd	n odd integer or n rational p odd, q odd	n rational p odd, q even	n irrational
Intercepts	(0,0,0)	(0,0,0)	(0,0,0)	(0,0,0)
Extrema	(0,0,0)	---	(0,0,0)	(0,0,0)
Inflection	---	(0,0,0)	---	---
Range	$-\infty < x < \infty$	$-\infty < x < \infty$	$0 \leq x < \infty$	$0 \leq x < \infty$
	$0 \leq y < \infty$	$-\infty < y < \infty$	$0 \leq y < \infty$	$0 \leq y < \infty$
Symmetry	$x = 0$	(0,0)	---	---
Cusp	(0,0) for n rational	---	---	---

Analysis of the Power Function.

$y = x^n$

$y' = nx^{n-1}$

$y'' = n(n-1) x^{n-2}$

$r = x\sqrt{1 + x^{2n-2}}$

$\tan \theta = x^{n-1}$

$m = nx^{n-1}$

$\tan \psi = \dfrac{(n-1) x^{n-1}}{1 + nx^{2n-2}}$

$p = -\dfrac{(n-1) x^n}{\sqrt{1 + n^2 x^{2n-2}}}$

$s' = \sqrt{1 + n^2 x^{2n-2}}$

$\rho = \dfrac{[1 + n^2 x^{2n-2}]^{3/2}}{n(n-1) x^{n-2}}$

$\alpha = \dfrac{x(n - n^2 x^{2n-2})}{n - 1}$

$\beta = x^n + \dfrac{1 + n^2 x^{2n-2}}{(n^2 - n) x^{n-2}}$

3.7. Polynomial

The polynomial is defined by

3.7.1) ... $\qquad y = \sum\limits_{i=0}^{n} a_i x^i$, $\quad a_0, \ldots, a_n$ constant .

Its general behavior near the origin depends on the value of the coefficients $a_0, \ldots, a_n$; far from the origin, the curve approaches $y = ax^n$.

Geometric properties depend on the coefficients, and the analytic properties are only an exercise in the summation notation. Polynomials are very important in many branches of mathematics; the theory of special curves is not, however, one of them.

CHAPTER 4

CUBIC CURVES

Quadratic curves are the only algebraic curves for which a complete analysis is generally known. Even the number of specific curves of degree higher than two is in dispute; cubic curves are classified into 57 to 219 varieties. (See references 5-7 for a detailed study of algebraic curves.)

Consequently, Chapters 4 through 6 describe only those curves that are widely known. Further, since these curves tend to be rather complicated, the section on analysis is generally abridged or omitted.

4.1. Semi-Cubical Parabola

The *semi-cubical parabola* (or *isochrone*) is defined to be the evolute of a parabola. If the parabola is given by

$$x = a(t^2 - 2)$$
$$y = -2 \, at$$

then the evolute (by equation 2.1.1) is

4.1.1) ... $\quad x = 3 \, at^2 \quad -\infty < t < \infty$.
$\quad\quad\quad\quad\quad y = 2 \, at^3$

Cubing and squaring, respectively, both sides of each equation in 4.2.1 yields the Cartesian equation,

4.1.2) ... $\quad 27 \, ay^2 = 4 \, x^3$.

From this equation, the polar form is easily found to be

4.1.3) ... $\quad 4 \, r = 27 \, a \, \sin^2\theta \, \sec^3\theta$.

Finally, an intrinsic equation is

$$s = 2a(\sec^3\psi - 1).$$

Examples are given in Figure 24.

The semi-cubical parabola may be generalized by defining it to be the curve represented by the equation

$$y^2 = ax^3 + bx^2 + cx + d.$$

A special case is Tschirnhausen's cubic,

$$27ay^2 = x^2(x + 9a);$$

see section 4.2.

Geometry of the Semi-Cubical Parabola.

Intercept	$(0, 0, 0)$
Extrema	$(0, 0, 0)$ is x-minimum
Extent	$-\infty < t < \infty$; $0 \le x < \infty$; $-\infty < y < \infty$
Symmetry	$y = 0$
Cusp of the first kind	$(0, 0, 0)$

Analysis of the Semi-Cubical Parabola.

$x = 3at^2$ $\qquad\qquad \tan\theta = \dfrac{2}{3}t$

$y = 2at^3$ $\qquad\qquad m = t$

$\dot{x} = 6at$ $\qquad\qquad \tan\psi = \dfrac{t}{3 + 2t^2}$

$\ddot{x} = 6a$

$\dot{y} = 6at^2$ $\qquad\qquad p = \dfrac{-at^3}{\sqrt{1 + t^2}}$

$\ddot{y} = 12at$

$y'' = \dfrac{1}{6at}$ $\qquad\qquad s = 2a[(1 + t^2)^{3/2} - 1]$

$\qquad\qquad\qquad\qquad \rho = 6at(1 + t^2)^{3/2}$

$r = at^2\sqrt{9 + 4t^2}$ $\qquad (\alpha, \beta) = at(-3t - 6t^3, 6 + 8t^2)$

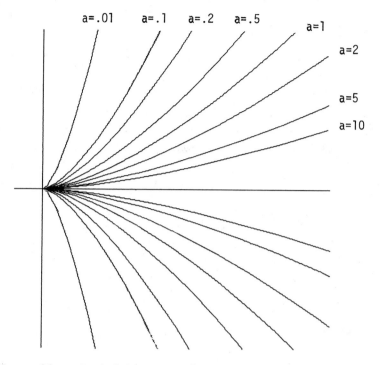

Figure 24.   Semi-Cubical Parabola
a = .01, .1, .2, .5, 1, 2, 5, 10

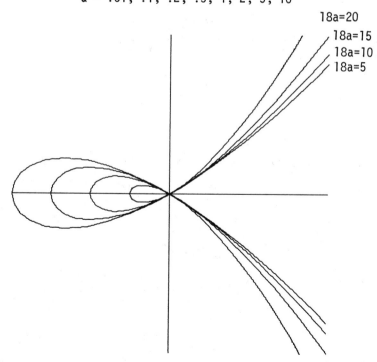

Figure 25.   Tschirnhausen's Cubic
18a = 5, 10, 15, 20

4.2. Tschirnhausen's Cubic

Tschirnhausen's cubic (also known as the *trisectrix of Catalan* and *l'Hospital's cubic*) is defined to be a Sinusoidal spiral (see section 7.1) with $n = -\frac{1}{3}$, given by

4.2.1) ... $\qquad r \cos^3 \frac{1}{3} \theta = a$.

If we define a parameter t by

$$t = \tan \frac{1}{3} \theta ,$$

the parametric equations

4.2.2) ... $\qquad \begin{cases} x = a (1 - 3 t^2) \\ y = a t (3 - t^2) \end{cases} \qquad -\infty < t < \infty$

may be easily derived. This may be put into a more regular form by a translation $x' = x + 8 a$, and a change of sign, to give

4.2.3) ... $\qquad \begin{cases} x = 3 a (t^2 - 3) \\ y = a t (t^2 - 3) \end{cases} \qquad -\infty < t < \infty$.

Eliminating t from these sets of equations quickly leads to the corresponding forms for the Cartesian equation,

4.2.4) ... $\qquad 27 a y^2 = (a - x) (x + 8 a)^2$

and

4.2.5) ... $\qquad 27 a y^2 = x^2 (x + 9 a)$,

respectively. The pedal equation for the first form is easily found to be

4.2.6) ... $\qquad a r^2 = p^3$.

It can be seen, from equations 4.2.4 and 4.2.5, that this cubic is one of the generalized semi-cubical parabolas discussed in section 4.1. It is illustrated in Figure 25.

The loop has an area of $\frac{72}{5} \sqrt{3} \, a^2$.

Geometry of Tschirnhausen's Cubic (for equation 4.2.3).

Intercepts	$(0, -9a, 0)$, $(\pm\sqrt{3}, 0, 0)$
Extrema	$(0, -9a, 0)$ is x-minimum
	$(\pm 1, -6a, \mp 2a)$ are y-minimum and maximum
Extent	$-\infty < t < \infty$
	$-9a \leq x < \infty$
	$-\infty < y < \infty$
Symmetry	$y = 0$
Loop	$\sqrt{3} \leq t \leq \sqrt{3}$
	$-9a \leq x \leq 0$
	$-2a \leq y \leq 2a$
Node	$(\pm\sqrt{3}, 0, 0)$

Analysis of Tschirnhausen's Cubic.

$x = 3a(t^2 - 3)$

$y = at(t^2 - 3)$

$\dot{x} = 6at$

$\ddot{x} = 6a$

$\dot{y} = 3a(t^2 - 1)$

$\ddot{y} = 6at$

$y'' = \dfrac{t^2 + 1}{12 a t^3}$

$r = a(t^2 - 3)\sqrt{9 + t^2}$

$$\tan \theta = t/3$$

$$m = \frac{t^2 - 1}{2t}$$

$$\tan \psi = \frac{t^2 - 3}{t(t^2 + 5)}$$

$$p = \frac{-a(t^2 - 3)^2}{t^2 + 1}$$

$$s = at(t^3 + 3)$$

$$\rho = \frac{3}{2}a(t^2 + 1)^2$$

$$\alpha = -\frac{3}{2}a(t^4 - 2t^2 + 5)$$

$$\beta = 4at^3$$

4.3. Witch of Agnesi (Fermat, 1666; Agnesi, 1748)

Consider a fixed circle C centered at (0,a) and tangent to the x-axis at the origin θ' (see the sketch below). For each secant through θ', let Q be the intersection of the secant and the circle; let A be the intersection of the secant and the line y = 2 a; and let P be the intersection of a line through Q parallel to the x-axis and a line through A parallel to the y-axis. The locus of P, for all such secants, is the *witch of Agnesi* (or *versiera*). It has one parameter, the radius, a, of the fixed circle. Examples are given in Figure 26.

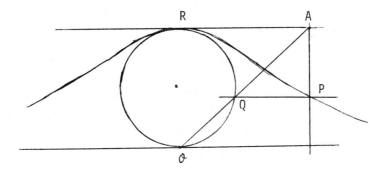

Let R be the point (0, 2 a), and let the line $\overline{OA}$ be given by $y = mx$, where $m = \cot t$ defines the parameter t. Since the circle is

$$x^2 + (y - a)^2 = a^2,$$

point Q is given by $Q = \dfrac{2am}{1 + m^2}(1, m)$. Now, $\overline{RA}$ and $\overline{OA}$ intersect at $A = 2a\left(\dfrac{1}{m}, 1\right)$. Hence, P is given by $P = 2a\left(\dfrac{1}{m}, \dfrac{m^2}{1 + m^2}\right)$.

From this, the parametric equation of the witch may be found:

4.3.1) ...
$$\begin{cases} x = 2a \tan t \\ y = 2a \cos^2 t \end{cases} \quad -\frac{\pi}{2} < t < \frac{\pi}{2}.$$

Solving equations 4.3.1 for t, and equating the results, yields the Cartesian equation

4.3.2) ... $$x^2 y = 4a^2 (2a - y);$$

substitution in this equation gives the polar equation

4.3.3) ... $$r^3 \sin^3\theta = r(r^2 + 4a^2) \sin\theta - 8a^3.$$

It is, perhaps, of interest to note that the area between the witch and the x-axis is equal to $4 \pi a^2$, and the volume of revolution of the witch about the x-axis is given by $V_x = 4 \pi^2 a^3$.

Geometry of the Witch of Agnesi.

Intercept	(0, 0, 2 a)
Extrema	(0, 0, 2 a) is y-maximum
Points of Inflection	$\left(\pm \frac{\pi}{6}, \pm \frac{\sqrt{3}}{3} a, \frac{3}{4} a \right)$
Extent	$-\pi < t < \pi; \; -\infty < x < \infty; \; 0 < y \leq 2a$
Symmetry	x = 0
Asymptote	y = 0

Analysis of the Witch of Agnesi.

$x = 2a \tan t$

$y = 2a \cos^2 t$

$\dot{x} = 2a \sec^2 t$

$\ddot{x} = 4a \sin t \sec^3 t$

$\dot{y} = -2a \sin 2t$

$\ddot{y} = -4a \cos 2t$

$y'' = \frac{1}{a} \cos^4 t \, (3 - 4\cos^2 t)$

$r = 2a \sqrt{\tan^2 t + \cos^4 t}$

$\tan \theta = \cos^3 t \csc t$

$m = -2 \sin t \cos^3 t$

$\tan \psi = \dfrac{(2\sin^2 t + 1) \cos^3 t}{(2\cos^6 t - 1) \sin t}$

$p = \dfrac{2a(1 + 2\sin^2 t)}{\sqrt{\sec^4 t + 4\sin^2 t \cos^2 t}}$

$\rho = \dfrac{a(1 + 4\cos^6 t \sin^2 t)^{3/2}}{(4\sin^2 t - 1) \cos^4 t}$

$\alpha = \dfrac{8a(1 + \cos^6 t) \sin^3 t}{(4\sin^2 t - 1) \cos t}$

$\beta = \dfrac{a(1 + 10\cos^6 t - 12\cos^8 t)}{(4\sin^2 t - 1) \cos^4 t}$

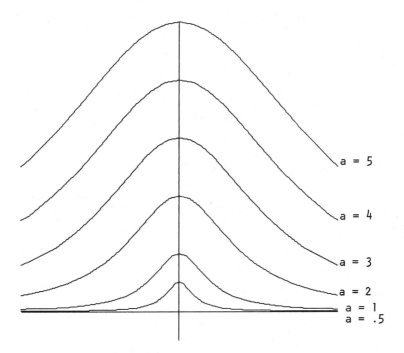

Figure 26. Witch of Agnesi
a = .5, 1, 2, 3, 4, 5

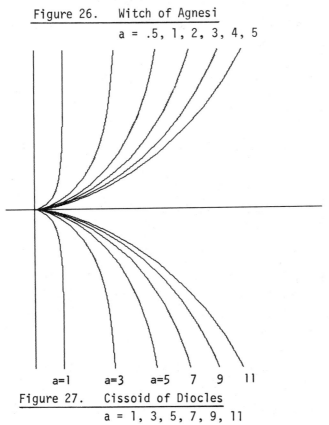

Figure 27. Cissoid of Diocles
a = 1, 3, 5, 7, 9, 11

4.4. Pedal of a Parabola

Consider the parabola $y^2 = -4a(x + am)$, or

4.4.1) ...
$$\begin{cases} f(t) = -at^2 - ma \\ g(t) = 2at \end{cases}$$

The pedal, with respect to (- am, 0), is given by equation 2.4.9; it is

4.4.2) ...
$$\begin{cases} x = a \dfrac{t^2 - m}{t^2 + 1} \\ y = at \dfrac{t^2 - m}{t^2 + 1} \end{cases}$$

Here, we have the parametric equation for a family of curves with an asymptotic and one loop. If we also replace t by tan t', the parametric equations become

4.4.3) ...
$$\begin{cases} x = a(\sin^2 t - m\cos^2 t) \\ y = a \tan t (\sin^2 t - m\cos^2 t) \end{cases} \quad -\frac{\pi}{2} < t < \frac{\pi}{2}.$$

The curve could be generalized by adding a multiplicative factor to the equation for y; this will not be explored here, however.

Solving the first equation of 4.4.3 for $\cos^2 t$ in terms of x, and substituting into the second equation, yields the Cartesian equation:

4.4.4) ...
$$y^2(a - x) = x^2(am + x).$$

Note that this introduces an isolated point (0, 0) that is not on 4.4.3 if m < 0. Special cases, and the sections in which they are discussed, are given in Table 5.

TABLE 5. (Generalized) Pedals of the Parabola

m	Section	Curve
0	4.5	Cissoid of Diocles
1	4.6	Right Strophoid
3	4.7	Trisectrix of Maclaurin

The area of the loop is $a^2 (m + 1) \left[\dfrac{\sqrt{m}}{2} \dfrac{m + 3}{m + 1} + \dfrac{m - 3}{2} \tan^{-1} \sqrt{m} \right]$.

The area between the curve and its asymptote is

$$a^2 (m + 1) \left[\frac{\pi}{4} (3 - m) + \frac{\sqrt{m}}{2} \frac{m + 3}{m + 1} + \frac{m - 3}{2} \tan^{-1} \sqrt{m} \right].$$

Examples, for various values of m, are given in Figures 27-30.

Geometry of the Pedal of a Parabola.

Intercepts $\qquad (0, - a m, 0), (\pm \tan^{-1} \sqrt{m}, 0, 0)$

Extrema $\qquad (0, - a m, 0)$ is x-minimum

$\qquad\qquad (\pm \hat{t}, \hat{x}, \pm \hat{y})$ are y-minimum,* maximum

where $\qquad \sin^2 \hat{t} = \dfrac{3}{4} - \dfrac{1}{4} \sqrt{\dfrac{m + 9}{m + 1}}$

$\qquad\qquad \hat{x} = \dfrac{1}{4} a \left(3 - m - \sqrt{(m + 9)(m + 1)} \right)$

Point of Inflection (for m < 0) $\quad \sin t = \pm \sqrt{\dfrac{m}{m - 3}}$

$\qquad\qquad x = \dfrac{4 \, a \, m}{m - 3} \, ; \, y = \dfrac{\pm 4 \, a \, m}{m - 3} \sqrt{-\dfrac{m}{3}}$

---

*y-minima exist only if m > 0.

Extent	$-\frac{\pi}{2} < t < \frac{\pi}{2}$
	$-am \leq x < a$ (for $m > -1$, $a > 0$; $m < -1$, $a < 0$)
	$a < x \leq -am$ (for $m > -1$, $a < 0$; $m < -1$, $a > 0$)
	$-\infty < y < \infty$
Symmetry	$y = 0$
Asymptote	$x = a$
Loop	$-\sqrt{m} \leq \tan(t) \leq \sqrt{m}$; $-am \leq x < 0$ for $m > 0$
Isolated Point	$(0, 0)$ for $m < 0$ (Cartesian Equation)
Node	$(\pm \tan^{-1}\sqrt{m}, 0, 0)$ for $m > 0$

Cusp of the first kind $(0, 0, 0)$ for $m = 0$.

Analysis of the Pedal of a Parabola.

$$x = a(\sin^2 t - m\cos^2 t)$$

$$y = a\tan t(\sin^2 t - m\cos^2 t)$$

$$\dot{x} = 2a(m+1)\sin t \cos t$$

$$\ddot{x} = 2a(m+1)(\cos^2 t - \sin^2 t)$$

$$\dot{y} = -a\sec^2 t [2(m+1)\sin^4 t - 3(m+1)\sin^2 t + m]$$

$$\ddot{y} = 2a\sin t \sec^3 t [2(m+1)\cos^4 t + 1]$$

$$y'' = \frac{m - (m-3)\sin^2 t}{4a(m+1)^2 \sin^3 t \cos^5 t}$$

$$r = a\sec t (\sin^2 t - m\cos^2 t)$$

$$\tan \theta = \tan t$$

$$m = -\frac{2(m+1)\sin^4 t - 3(m+1)\sin^2 t + m}{2(m+1)\sin t \cos^3 t}$$

$$\tan \psi = \cot t \frac{(m+1)\sin^2 t - m}{m + 2 - (m+1)\sin^2 t}$$

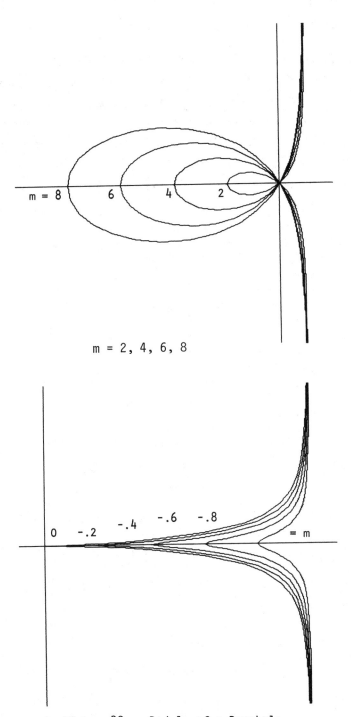

m = 2, 4, 6, 8

Figure 28. Pedals of a Parabola
a = 1
m = 2, 4, 6, 8; -.8, -.6, -.4, -.2, 0

4.5. Cissoid of Diocles (Diocles, ca. 200 B.C.)

The *cissoid of Diocles* is defined to be the cissoid of a circle and a tangent line, with respect to a fixed point on the circumference opposite the point of tangency. It has one parameter, the diameter, a, of the circle. If the fixed point is on the circumference, but is not opposite the point of tangency, the curve is termed an oblique cissoid. Examples are given in Figure 27. The derivation below refers to the sketch

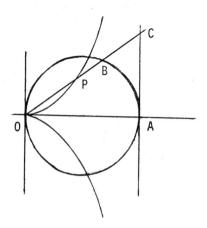

Let $x = a$ be the line tangent to the circle $(x - \frac{a}{2})^2 + y^2 = \frac{1}{4} a^2$. Then, the fixed point is the origin $(0, 0)$. Now, $\overline{OP} = \overline{OC} - \overline{OB}$, so, if $P = (x, y)$ is on the cissoid, we have

$$C = a (1, \tan \theta)$$
$$B = a \cos \theta (\cos \theta, \sin \theta) .$$

Hence, $r = \overline{OP} = a \sec \theta - a \cos \theta = a \sin \theta \tan \theta$, so the polar equation of the cissoid is

4.5.1) ... $\qquad r = a \sin \theta \tan \theta$.

The Cartesian equation follows immediately by substitution,

4.5.2) ... $\qquad y^2 (a - x) = x^3$.

If we now set $\tan t = \tan \theta$, we may use 4.5.1 to derive

$$x = r \cos \theta = a \sin^2 t$$

and

$$y = r \sin \theta = a \tan t \sin^2 t .$$

Hence, the parametric equations of the cissoid are

4.5.3) ... $$\begin{cases} x = a \sin^2 t \\ y = a \tan t \sin^2 t \end{cases} \quad -\infty < t < \infty .$$

The Whewell and Cesáro intrinsic equations are

$$s = a (\sec^2 \phi - 1)$$

and

$$729 (s + a)^8 = a^2 [9 (s + a)^2 + \rho^2]^3 ,$$

respectively.

The volume of revolution about the asymptote is given by $V = \frac{1}{4} \pi^2 a^3$. The area between the cissoid and the asymptote is equal to $\frac{3}{4} \pi a^2$.

Geometry and analysis can be found in section 4.4, using $m = 0$. For convenience, the analysis is also displayed in this section.

Analysis of the Cissoid of Diocles.

$x \quad = a \sin^2 t$

$y \quad = a \tan t \sin^2 t$

$\dot{x} \quad = 2 a \sin t \cos t$

$\ddot{x} \quad = 2 a (\cos^2 t - \sin^2 t)$

$\dot{y} \quad = a \tan^2 t (1 + 2 \cos^2 t)$

$\ddot{y} \quad = 2 a \tan t \sec^2 t (1 + 2 \cos^4 t)$

$y'' \quad = \dfrac{3}{4 a \sin t \cos^5 t}$

$$r = a \sin^2 t \sec t$$

$$\tan \theta = \tan t$$

$$m = \frac{1}{2} \tan t \sec^2 t \, (1 + 2 \cos^2 t)$$

$$\tan \psi = \frac{\sin t \cos t}{1 + \cos^2 t}$$

$$p = -\frac{a \sin^3 t}{\sqrt{1 + 3 \cos^2 t}}$$

$$\rho = \frac{1}{6} a \sin t \sec^4 t \, [1 + 3 \cos^2 t]^{3/2}$$

$$\alpha = -\frac{1}{6} a \sin^2 t \sec^4 t \, (1 + 5 \cos^2 t)$$

$$\beta = \frac{4}{3} a \tan t$$

4.6. Right Strophoid (Barrow, 1670)

The *right strophoid* is defined to be the strophoid of a line $\mathcal{L}$ with respect to two points A and $\mathcal{O}$, where A is the foot of the perpendicular from the pole $\mathcal{O}$ to the line. If A is on $\mathcal{L}$, but is not the foot of the perpendicular, then the strophoid is termed an *oblique strophoid*.

The curve may also be defined as the pedal of a parabola with respect to the intersection of the axis and the directrix.

The equation for the right strophoid is derived easiest by using the polar equation of the strophoid 2.6.8. Let the pole $\mathcal{O}$ be at the origin, and let $A = (a, 0)$ (so $r_0 = a$, $\theta_0 = 0$). Then the line has polar equation $r = a \sec \theta$ (from 1.1.8). Substitution in 2.6.8 yields the polar equation of the right strophoid,

4.6.1) ... $\quad r = a \, (\sec \theta \pm \tan \theta)$.

Standard substitutions for x and y give the Cartesian equation,

4.6.2) ... $$x(x - a)^2 = y^2 (2a - x).$$

The more usual form of the strophoid has the fixed point at the origin and the pole at $(-a, 0)$. This is, of course, just a translation along the x-axis, which transforms 4.6.2 to

4.6.3) ... $$y^2 (a - x) = x^2 (a + x).$$

Transforming this back to polar form gives

4.6.4) ... $$r = a(\sec\theta - 2\cos\theta).$$

This last equation may be easily transformed to parametric form by substituting the value of r in 4.6.4 into the equations $x = r\cos\theta$ and $y = r\sin\theta$. This gives

$$x = a(1 - 2\cos^2\theta)$$
$$y = a(\tan\theta - 2\sin\theta\cos\theta).$$

Letting $t = \theta$, we have the parametric equations

4.6.5) ... $$\begin{cases} x = a(1 - 2\cos^2 t) \\ y = a\tan t\,(1 - 2\cos^2 t). \end{cases}$$

Translating this back along the x-axis finally gives the parametric equation corresponding to 4.6.1 and 4.6.2:

4.6.6) ... $$\begin{cases} x = 2a\cos^2 t \\ y = a\tan t\,(1 - 2\cos^2 t). \end{cases}$$

In the case of the oblique strophoid, the defining line is $r = a\cos\alpha\,\sec(\theta - \alpha)$. Using 2.6.8 gives (after some manipulation)

4.6.7) ... $$r = a(\cos\alpha \pm \sin\theta)\sec(\theta - \alpha);$$

it will be noted that this reduces to 4.6.1 when $\alpha = 0$. As this leads to a fourth degree equation, it will not be explored further.

The table below summarizes the equations for the two forms of the right strophoid. The curve is illustrated in Figure 29.

	Form-a	Form-b
O	$(-a, 0)$	$(0, 0)$
A	$(0, 0)$	$(a, 0)$
line	$x = 0$	$x = a$
Cartesian	$y^2 (a - x) = x^2 (a + x)$	$x (x - a)^2 = y^2 (2a - x)$
Polar	$r = a (\sec \theta - 2 \cos \theta)$	$r = a (\sec \theta \pm \tan \theta)$
Parametric	$\begin{cases} x = a (1 - 2 \cos^2 t) \\ y = a \tan t (1 - 2 \cos^2 t) \end{cases}$	$x = 2a \cos^2 t$ $y = a \tan t (1 - 2 \cos^2 t)$

Finally, some area formulae. The area of the loop is $a^2 (2 - \pi/2)$ and the area between the curve and its asymptote is $a^2 (2 + \pi/2)$. The geometry and analysis can be found in section 4.4, with $m = 1$.

Analysis of the Right Strophoid (Form-a).

$x = a (1 - 2 \cos^2 t)$

$y = a \tan t (1 - 2 \cos^2 t)$

$\dot{x} = 4 a \sin t \cos t$

$\ddot{x} = 4 a (2 \cos^2 t - 1)$

$\dot{y} = - a \sec^2 t (4 \sin^4 t - 6 \sin^2 t + 1)$

$\ddot{y} = 2 a \sin t \sec^3 t (4 \cos^4 t + 1)$

$y'' = \dfrac{1 + 2 \sin^2 t}{16 a \sin^3 t \cos^5 t}$

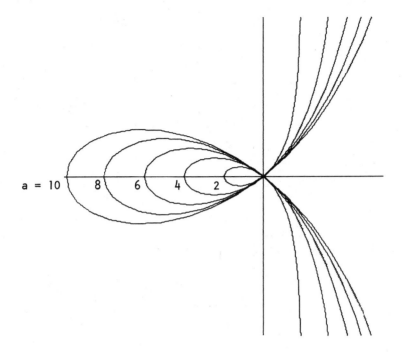

Figure 29. Right Strophoid
a = 2, 4, 6, 8, 10

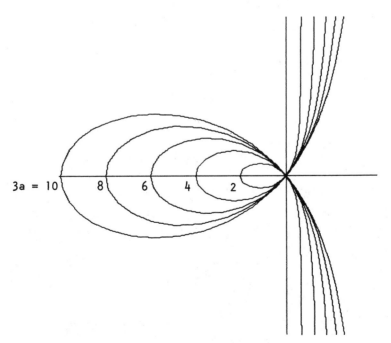

Figure 30. Trisectrix of Maclaurin
3a = 2, 4, 6, 8, 10

$$r = a \sec t\, (1 - 2\cos^2 t)$$

$$\tan\theta = \tan t$$

$$m = -\frac{4\sin^4 t - 6\sin^2 t + 1}{4\sin t \cos^3 t}$$

$$\tan\psi = \cot t\, \frac{1 - 2\cos^2 t}{1 + 2\cos^2 t}$$

$$p = \frac{-a(1 - 2\cos^2 t)^2}{\sqrt{1 + 4\cos^2 t - 4\cos^4 t}}$$

$$\rho = \frac{a[1 + 4\cos^2 t - 4\cos^4 t]^{3/2}}{4\cos^4 t\,(1 + 2\sin^2 t)}$$

$$\alpha = \frac{a(8\sin^6 t - 12\sin^4 t + 6\sin^2 t - 3)}{4\cos^4 t\,(1 + 2\sin^2 t)}$$

$$\beta = \frac{4a\sin^3 t}{\cos t\,(1 + 2\sin^2 t)}$$

4.7. Trisectrix of Maclaurin (Maclaurin, 1742)

The *trisectrix of Maclaurin* may be defined as the pedal of a parabola with respect to the reflection of the focus on the directrix. The methods of section 4.4 with m = 3 lead immediately to the equation for the trisectrix,

4.7.1) ...
$$\begin{cases} x = a\,\dfrac{t^2 - 3}{t^2 + 1} \\[2mm] y = a t\,\dfrac{t^2 - 3}{t^2 + 1} \end{cases} \qquad -\infty < t < \infty.$$

The Cartesian equation is

4.7.2) ... $$y^2(a-x) = x^2(x+3a)$$

and the corresponding polar form is

4.7.3) ... $$r = a\sec\theta - 4a\cos\theta.$$

The loop has an area of $3\sqrt{3}\, a^2$.

A more convenient parametric form is found by replacing t (in 4.7.1) by tan t'. This results in

4.7.4) ... $$\begin{cases} x = a(1 - 4\cos^2 t) \\ y = a\tan t\,(1 - 4\cos^2 t) \end{cases} \quad -\frac{\pi}{2} < t < \frac{\pi}{2}.$$

The geometry and analysis for equations 4.7.1 may be obtained from section 4.4, using m = 3. The facts for equations 4.7.4 are given in this section. The curve is illustrated in Figure 30.

Geometry of the Trisectrix of Maclaurin.

Intercepts	$(\pm\frac{\pi}{3}, 0, 0)$, $(0, -3a, 0)$
Extrema	$(0, -3a, 0)$ is x-minimum
Extent	$-\frac{\pi}{2} < t < \frac{\pi}{2}$; $-3a \leq x < a$; $-\infty < y < \infty$
Symmetry	$y = 0$
Asymptote	$x = a$
Loop	$-\frac{\pi}{3} \leq t \leq \frac{\pi}{3}$; $-3a \leq x \leq 0$
Node	$(\pm\frac{\pi}{3}, 0, 0)$

Analysis of the Trisectrix of Maclaurin.

$x \quad = a(1 - 4\cos^2 t)$

$y \quad = a\tan t\,(1 - 4\cos^2 t)$

$\dot{x} \quad = 8a\sin t\cos t$

$$\ddot{x} = 8a(\cos^2 t - \sin^2 t)$$

$$\dot{y} = a\sec^2 t(-8\sin^4 t + 12\sin^2 t - 3)$$

$$\ddot{y} = 2a\sin t(\sec^3 t + 8\cos t)$$

$$y'' = \frac{3}{64\,a}\csc^3 t \sec^5 t$$

$$r = a(\sec t - 4\cos t)$$

$$\theta = t$$

$$m = \frac{-8\sin^4 t + 12\sin^2 t - 3}{8\sin t \cos^3 t}$$

$$\tan\psi = \cot t\,\frac{1 - 4\cos^2 t}{1 + 4\cos^2 t}$$

$$p = \frac{-a(1 - 4\cos^2 t)^2}{\sqrt{9 - 8\sin^2 t}}$$

$$\rho = \frac{1}{24}a(9 - 8\sin^2 t)^{3/2} \sec^4 t$$

$$\alpha = -\frac{1}{24}a\sec^4 t(32\cos^6 t + 12\cos^2 t + 1)$$

$$\beta = \frac{4}{3}a\sin^3 t \sec t$$

4.8. Folium of Descartes (Descartes, 1638)

The *folium of Descartes* is the curve described by the equation

4.8.1) ... $\qquad x^3 + y^3 = 3axy$.

Clearly, the polar form is

4.8.2) ... $\qquad r(\sin^3\theta + \cos^3\theta) = 3a\sin\theta\cos\theta$.

The parametric representation is found by letting y = x t in 4.8.1, yielding

4.8.3) ...
$$\begin{cases} x = \dfrac{3\,a\,t}{1+t^3} \\[2mm] y = \dfrac{3\,a\,t^2}{1+t^3} \end{cases} \qquad -\infty < t < \infty.$$

In this form, the curve has three arcs (see Figure 31). For $-1 < t < 0$, the curve is located in the second quadrant, with $t = 0$ corresponding to the origin. For $t < -1$, the curve occupies the fourth quadrant, and approaches the origin as $t \to -\infty$. The loop in the first quadrant corresponds to $0 \leq t < \infty$, going counterclockwise with increasing t.

It is interesting to rotate the curve through an angle $\theta = \frac{5}{4}\pi$. If we also replace a by $\sqrt{2}\,a$, this yields

4.8.4) ... $\qquad 3\,y^2\,(a - x) = x^2\,(x + 3\,a),$

with polar equation

4.8.5) ... $\qquad 3\,\tan^2\theta = \dfrac{3\,a + r\,\cos\theta}{a - r\,\cos\theta}$

and parametric equations

4.8.6) ...
$$\begin{cases} x = 3\,a\,\dfrac{t^2 - 1}{3\,t^2 + 1} \\[2mm] y = 3\,a\,t\,\dfrac{t^2 - 1}{3\,t^2 + 1} \end{cases}.$$

In this form, it can be seen that the curve is related to the curves discussed in section 4.4.

The loop of the folium has area $\frac{3}{2}\,a^2$, as does the region between the curve and its asymptote $x + y + a = 0$.

Geometry of the Folium of Descartes.

Intercept	$(0, 0, 0)$
Extrema	$(0, 0, 0)$ is y-minimum and x-minimum
	$(2^{1/3}, 2^{1/3}a, 2^{2/3}a)$ is y-maximum
	$(2^{-1/3}, 2^{2/3}a, 2^{1/3}a)$ is x-maximum
Extent	$-\infty < t < \infty$
	$-\infty < x < \infty$
	$-\infty < y < \infty$
Discontinuity	for $t = -1$
Symmetry	$x = y$
Asymptote	$x + y + a = 0$
Loop	$0 \leq t < \infty$
Node	$(0, 0, 0), (+\infty, 0, 0), (-\infty, 0, 0)$

Analysis of the Folium of Descartes.

$$x = \frac{3 a t}{1 + t^3}$$

$$\ddot{y} = 6 a \frac{1 - 7 t^3 + t^6}{(1 + t^3)^3}$$

$$y = \frac{3 a t^2}{1 + t^3}$$

$$y'' = \frac{2 (1 + t^3)^4}{3 a (1 - 2 t^3)^3}$$

$$\dot{x} = \frac{3 a (1 - 2 t^3)}{(1 + t^3)^2}$$

$$r = \frac{3 a t}{1 + t^3} \sqrt{1 + t^2}$$

$$\ddot{x} = 18 a t^2 \frac{t^3 - 2}{(1 + t^3)^3}$$

$$\tan \theta = t$$

$$\dot{y} = 3 a t \frac{2 - t^3}{(1 + t^3)^2}$$

$$m = \frac{t (2 - t^3)}{1 - 2 t^3}$$

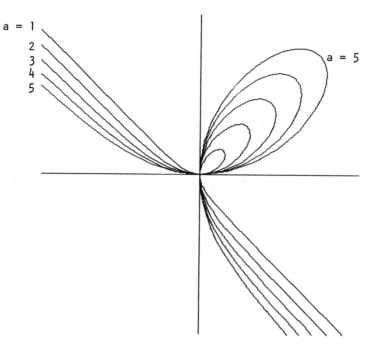

Figure 31. Folium of Descartes
a = 1, 2, 3, 4, 5

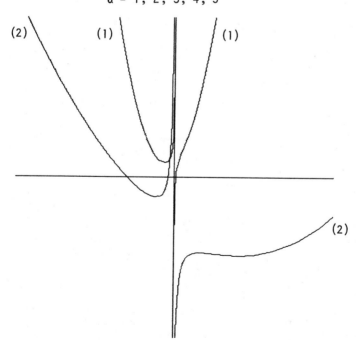

Figure 32. Trident of Newton
1) a = 1, b = 1, c = 1, d = -.1
2) a = .0762, b = -.686, c = -3.105, d = -1.143

4.9. Trident of Newton (Newton, 1701)

The *trident of Newton*, also knows as the *parabola of Descartes* (Figure 32), is defined by the equation

4.9.1) ... $\quad x y = a x^3 + b x^2 + c x + d , \quad a d \neq 0 .$

Neither parametric nor polar equations are interesting. In the discussion below, $u = a x^3 + b x^2 + c x + d$.

Geometry of the Trident of Newton.

Intercepts   given by the real roots of u.
Discontinuity $x = 0$ (infinite).
Asymptote $x = 0$.

Analysis of the Trident of Newton.

$$y = \frac{a x^3 + b x^2 + c x + d}{x}$$

$$m = y' = \frac{2 a x^3 + b x^2 - d}{x^2}$$

$$y'' = \frac{2 (a x^3 + d)}{x^3}$$

$$r = \frac{1}{x} \sqrt{x^4 + u^2}$$

$$\tan \theta = \frac{u}{x^2}$$

4.10. Serpentine (Newton, 1701)

The *serpentine* is the curve defined by the Cartesian equation

4.10.1) ... $\quad x^2 y + a^2 y - b^2 x = 0$,

and is a projection of the *horopter*, the intersection of a cylinder and a hyperbolic paraboloid.

As 4.10.1 is solvable for y directly, there is no necessity for a parametric form. The polar equation is

4.10.2) ... $\quad r^2 \cos^2\theta = b^2 \cot\theta - a^2$.

The serpentine is illustrated in Figure 33.

Geometry of the Serpentine.

Intercept	$(0, 0)$
Extrema	$\left(\pm a, \pm \dfrac{b^2}{2a}\right)$ for y-maximum, minimum
Inflection	$x = 0; \quad x = \pm \dfrac{\sqrt{3}}{3} a$
Extent	$-\infty < x < \infty$
	$-\dfrac{b^2}{2a} < y < \dfrac{b^2}{2a}$
Symmetry	$(0, 0)$
Asymptote	$y = 0$

Analysis of the Serpentine.

$y = \dfrac{b^2 x}{x^2 + a^2}$ $\qquad r = \dfrac{x}{x^2 + a^2} \sqrt{b^4 + (x^2 + a^2)^2}$

$y' = \dfrac{b^2 (a^2 - x^2)}{(x^2 + a^2)^2}$ $\qquad \tan\theta = \dfrac{b^2}{x^2 + a^2}$

$y'' = \dfrac{2 b^2 x (x^2 - 3 a^2)}{(x^2 + a^2)^3}$ $\qquad m = y'$

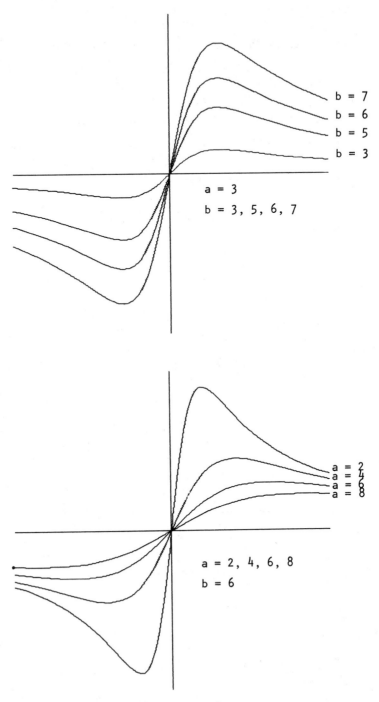

Figure 33. Serpentine
 a = 3 b = 3, 5, 6, 7
 a = 2, 4, 6, 8 b = 6

CHAPTER 5

QUARTIC CURVES

5.1. Limacon of Pascal (Pascal, 1650)

The *limacon of Pascal* (Figure 34) may be defined geometrically as the conchoid of a circle of radius a, with respect to a point O on the circumference of the circle. If we let O be the origin, and define the circle by $r = 2 a \cos \theta$, then, by equation 2.5.10, the limacon is given by

5.1.1) ... $\qquad r = 2 a \cos \theta + b$,

where b is the parameter of the conchoid.

There is another definition of the limacon frequently used. It is the epitrochoid (roulette) generated by a point rigidly attached to a circle rolling upon an equal fixed circle. This will be discussed in Chapter 7.

If $b = 2a$, the limacon becomes a cardioid (see section 5.2). If $b = a$, it is a trisectrix (Figure 35). If $a = 0$, it is a circle.

The parametric equations may be found from 5.1.1 by using the equations $x = r \cos \theta$ and $y = r \sin \theta$, yielding

5.1.2) ... $\qquad \begin{cases} x = a \cos 2\theta + b \cos \theta + a \\ y = a \sin 2\theta + b \sin \theta. \end{cases}$

If this is expanded, and t is used instead of θ, the equations

5.1.3) ... $\qquad \begin{cases} x = \cos t \, (2 a \cos t + b) \\ y = \sin t \, (2 a \cos t + b) \end{cases} \qquad -\pi \leq t \leq \pi$

result.

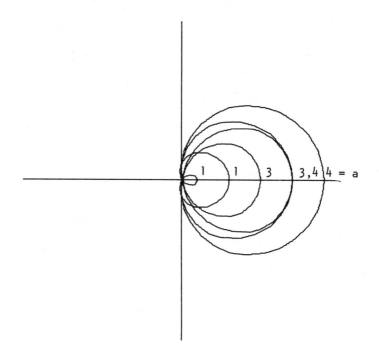

Figure 34. Limacon of Pascal
a = 1, 3, 4; b = 1

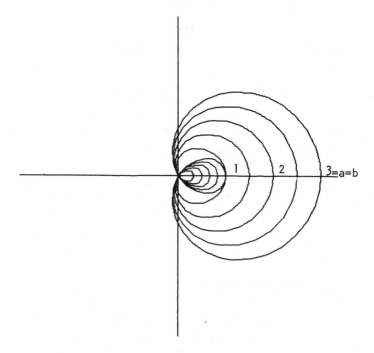

Figure 35. Trisectrix
a = b = 1, 1.5, 2, 2.5, 3

The Cartesian equation follows quickly from 5.1.1:

$$r - 2a\cos\theta = b$$

$$r^2 - 2ar\cos\theta = br \quad \text{(introducing an isolated point at the origin)}$$

$$(r^2 - 2ar\cos\theta)^2 = b^2 r^2$$

or

5.1.4) ... $\qquad (x^2 + y^2 - 2ax)^2 = b^2(x^2 + y^2)$.

The limacon forms a single loop if $b \geq 2a$, and two loops if $b < 2a$. In the former case, the loop has an area of $(2a^2 + b^2)\pi$. In the latter case, the outer loop has area

$$(2a^2 + b^2)\left(\pi - \cos^{-1}\frac{b}{2a}\right) + \frac{3}{2}b\sqrt{4a^2 - b^2}$$

and the inner loop has area

$$(2a^2 + b^2)\cos^{-1}\frac{b}{2a} - \frac{3}{2}b\sqrt{4a^2 - b^2} .$$

If $b \geq 2a$, the limacon has a surface of revolution

$$\Sigma_x = \frac{2\pi}{15 a b^2}\left[(2a+b)^4(a-2b) - (2a-b)^4(a+2b)\right]$$

and a volume of revolution

$$V_x = \frac{2}{3}a\pi(4a^2 + b) .$$

Geometry of the Limacon, $b > 2a$.

Intercepts $\qquad (\pm\pi, 2a - b, 0), \; (\pm\frac{\pi}{2}, 0, \pm b), \; (0, 2a + b, 0)$

Extrema $\qquad (0, 2a + b, 0)$ is x-maximum

$$\left(\pm\pi \mp \cos^{-1}\frac{b}{4a}, \; -\frac{b^2}{4a}, \; \pm\frac{b}{8a}\sqrt{16a^2 - b^2}\right)$$

are x-minima if $2a \leq b \leq 4a$

y-extrema are for $\cos t = \frac{1}{8a}\left(-b \pm \sqrt{b^2 + 32a^2}\right)$.

Extent	$-\pi \leq t \leq \pi$
	$-\dfrac{b^2}{8a} \leq x \leq 2a + b$
Symmetry	$y = 0$
Loop	Entire curve
Isolated Point	$\left(\pm \pi \mp \cos^{-1}\dfrac{b}{2a},\ 0,\ 0\right)$ for equation 5.1.4 only

Geometry of the Limacon, $b < 2a$.

Intercepts	$(\pm \pi,\ 2a - b,\ 0)$, $\left(\pm \dfrac{\pi}{2},\ 0,\ \pm b\right)$, $(0,\ 2a + b,\ 0)$,
	$\left(\pm \pi \mp \cos^{-1}\dfrac{b}{2a},\ 0,\ 0\right)$
Extrema	$(0,\ 2a + b,\ 0)$ is x-maximum
	$\left(\pm \pi \mp \cos^{-1}\dfrac{b}{4a},\ -\dfrac{b^2}{4a},\ \pm \dfrac{b}{8a}\sqrt{16a^2 - b^2}\right)$ are x-minima
	y-extrema are for $\cos t = \dfrac{1}{8a}\left(-b \pm \sqrt{b^2 + 32a^2}\right)$
Extent	$-\pi \leq t \leq \pi$
	$-\dfrac{b^2}{8a} \leq x \leq 2a + b$
Symmetry	$y = 0$
Loops	a) $-\pi + \cos^{-1}\dfrac{b}{4a} \leq t \leq \pi - \cos^{-1}\dfrac{b}{4a}$ is the outer loop
	b) $-\pi \leq t \leq -\pi + \cos^{-1}\dfrac{b}{4a}$; $\pi - \cos^{-1}\dfrac{b}{4a} \leq t \leq \pi$
	is the inner loop
Node	$\left(\pm \pi \mp \cos^{-1}\dfrac{b}{2a},\ 0,\ 0\right)$

Analysis of the Limacon of Pascal.

$x = a \cos 2t + b \cos t = \cos t (2a \cos t + b)$

$y = a \sin 2t + b \sin t = \sin t (2a \cos t + b)$

$\dot{x} = -2a \sin 2t - b \sin t = -\sin t (4a \cos t + b)$

$\ddot{x} = -4a \cos 2t - b \cos t = -(8a \cos^2 t + b \cos t - 4a)$

$\dot{y} = 2a \cos 2t + b \cos t = 4a \cos^2 t + b \cos t - 2a$

$\ddot{y} = -4a \sin 2t - b \sin t = -\sin t (8a \cos t + b)$

$y'' = -\dfrac{6ab \cos t + 8a^2 + b^2}{\sin^3 t (4a \cos t + b)^3}$

$r = 2a \cos t + b$

$\theta = t$

$m = -\dfrac{4a \cos^2 t + b \cos t - 2a}{\sin t (4a \cos t + b)}$

$\tan \psi = -\dfrac{2a \cos t + b}{2a \sin t}$

$p = -\dfrac{(2a \cos t + b)^2}{\sqrt{4ab \cos t + 4a^2 + b^2}}$

$\rho = \dfrac{[4ab \cos t + 4a^2 + b^2]^{3/2}}{6ab \cos t + 8a^2 + b^2}$

5.2. Cardioid (Koërsma, 1689)

The *cardioid* (Figure 36) is a limacon for which $b = 2a$; its equations, therefore, are:

5.2.1) ... $\quad r = 2a(1 + \cos \theta)$

5.2.2) ... $\quad \begin{cases} x = 2a \cos t\,(1 + \cos t) \\ y = 2a \sin t\,(1 + \cos t) \end{cases} \quad -\pi \leq t \leq \pi.$

5.2.3) ... $\quad (x^2 + y^2 - 2ax)^2 = 4a^2(x^2 + y^2).$

The pedal equation is

5.2.4) ... $\quad r^3 = 4ap^2 \quad \text{or} \quad 9(r^2 - a^2) = 8p^2$

and the intrinsic equations are

5.2.5) ... $\quad s^2 + 9\rho^2 = 64a^2 \quad \text{(Cesáro)}$

5.2.6) ... $\quad s = 8a\left(1 - \cos \frac{1}{3}\psi\right), \quad 0 \leq \psi \leq 3\pi$

5.2.7) ... $\quad s = 8a \cos \frac{1}{3}\phi \quad \text{(Whewell)}.$

Interestingly, its evolute is another cardioid, with parameter $a' = \frac{1}{3}a$.

The loop has area $6\pi a^2$ and length $16a$; the surface area and volume of revolution are

$$\Sigma_x = \frac{128}{5}\pi a^2 \quad \text{and} \quad V_x = \frac{4}{3}a^2 \pi (2a + 1).$$

Geometry of the Cardioid.

Intercepts $\quad (\pm \pi, 0, 0), \quad \left(\pm \frac{\pi}{2}, 0, \pm 2a\right), \quad (0, 4a, 0)$

Extrema $\quad (0, 4a, 0)$ is x-maximum

$\quad \left(\pm \frac{2}{3}\pi, -a, \pm \frac{1}{2}\sqrt{3}\,a\right)$ are x-minima

$\quad \left(\pm \frac{\pi}{3}, \frac{3}{2}a, \pm \frac{3\sqrt{3}}{2}a\right)$ are y-extrema

Extent	$-\pi \le t \le \pi$
	$-a \le x \le 4a$
	$-\frac{3\sqrt{3}}{2} a \le y \le \frac{3\sqrt{3}}{2} a$
Symmetry	$y = 0$
Loop	entire curve
Cusp	$(\pm \pi, 0, 0)$

Analysis of the Cardioid.

$x = 2a \cos t (1 + \cos t)$

$y = 2a \sin t (1 + \cos t)$

$\dot{x} = -2a \sin t (1 + 2 \cos t)$

$\ddot{x} = -2a (4 \cos^2 t + \cos t - 2)$

$\dot{y} = 2a (2 \cos t - 1)(\cos t + 1)$

$\ddot{y} = -2a \sin t (1 + 4 \cos t)$

$y'' = -\dfrac{3 (1 + \cos t)}{2a \sin^3 t (1 + 2 \cos t)^3}$

$r = 2a (1 + \cos t)$

$\theta = t$

$m = -\dfrac{(2 \cos t - 1)(\cos t + 1)}{\sin t (2 \cos t + 1)}$

$\tan \psi = -\dfrac{1 + \cos t}{\sin t}$

$p = \sqrt{2} a (1 + \cos t)^{3/2}$

$s = 8a \sin \frac{1}{2} t$

$\rho = \frac{4}{3}\sqrt{2}\, a\sqrt{1 + \cos t}$

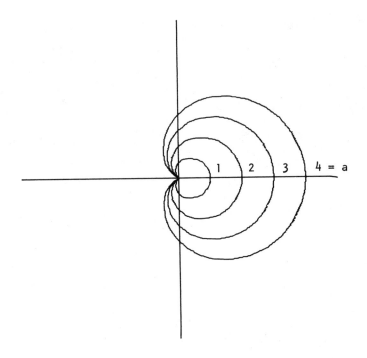

Figure 36. Cardioid
a = 1, 2, 3, 4

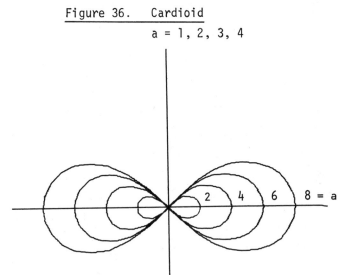

Figure 37. Lemniscate of Bernoulli
a = 2, 4, 6, 8

$$\alpha = \frac{2}{3} a (2 - \cos t)(1 + \cos t)$$

$$\beta = \frac{2}{3} a \sin t (1 - \cos t)$$

5.3. Lemniscate of Bernoulli (Bernoulli, 1694)

The *lemniscate of Bernoulli*, also termed the *hyperbolic lemniscate*, is defined as the sinusoidal spiral for which n = 2, and as the inverse of an equilateral hyperbola, with respect to the origin. If the hyperbola is

$$f(t) = a \sec t$$
$$g(t) = a \tan t ,$$

then its inverse (by 2.3.9, with $k = a^2$) is

$$x = \frac{a^2 f(t)}{f^2(t) + g^2(t)} = \frac{a^3 \sec t}{a^2 (\sec^2 t + \tan^2 t)}$$

$$y = \frac{a^2 g(t)}{f^2(t) + g^2(t)} = \frac{a^3 \tan t}{a^2 (\sec^2 t + \tan^2 t)} .$$

Hence, the parametric equations of the lemniscate are

5.3.1) ... $\begin{cases} x = \dfrac{a \cos t}{1 + \sin^2 t} \\ \\ y = \dfrac{a \sin t \cos t}{1 + \sin^2 t} \end{cases}$ $-\pi \leq t \leq +\pi$.

Now,

$$x^2 + y^2 = a^2 \frac{\cos^2 t}{1 + \sin^2 t}$$

and

$$x^2 - y^2 = a^2 \frac{\cos^4 t}{(1 + \sin^2 t)^2} .$$

Hence, the Cartesian equation is

5.3.2) ... $(x^2 + y^2)^2 = a^2 (x^2 - y^2)$.

Substitution immediately leads to

5.3.3) ... $r^2 = a^2 \cos 2\theta$,

the polar equation and the sinusoidal spiral form; (see section 7.1). The pedal and bipolar equations are

5.3.4) ... $r^3 = a^2 p$

5.3.5) ... $r \, r' = \frac{1}{2} a^2$.

The lemniscate is illustrated in Figure 37. The area of the loops is $A = a^2$, and the volume of revolution is $V_x = 2 \pi a^2 (2 - \sqrt{2})$.

Geometry of the Lemniscate.

Intercepts $(0, a, 0)$, $(\pm \frac{\pi}{2}, 0, 0)$, $(\pm \pi, -a, 0)$

Extrema $(0, a, 0)$ is x-maximum

$(\pm \pi, -a, 0)$ is x-minimum

$\left(\sin^{-1} \frac{\sqrt{3}}{3}, \pm \frac{a\sqrt{6}}{4}, \pm \frac{a\sqrt{2}}{4} \right)$ are y-extrema

Inflection $\left(\pm \frac{\pi}{2}, 0, 0 \right)$

Extent $-\pi \le t \le \pi$

$-a \le x \le a$

$-\frac{a\sqrt{2}}{8} \le y \le \frac{a\sqrt{2}}{8}$

Symmetries $x = 0$; $y = 0$; $(0, 0)$

Loops   two: 1) $t \le -\frac{\pi}{2}$ and $t \ge \frac{\pi}{2}$

2) $-\frac{\pi}{2} \le t \le \frac{\pi}{2}$

Node $\left(\pm \frac{\pi}{2}, 0, 0 \right)$

Analysis of the Lemniscate.

$$x = \frac{a \cos t}{1 + \sin^2 t}$$

$$y = \frac{a \sin t \cos t}{1 + \sin^2 t}$$

$$\dot{x} = -a \sin t \frac{3 - \sin^2 t}{(1 + \sin^2 t)^2}$$

$$\ddot{x} = -a \cos t \frac{3 - 12 \sin^2 t + \sin^4 t}{(1 + \sin^2 t)^3}$$

$$\dot{y} = a \frac{1 - 3 \sin^2 t}{(1 + \sin^2 t)^2}$$

$$\ddot{y} = 2a \sin t \cos t \frac{3 \sin^2 t - 5}{(1 + \sin^2 t)^3}$$

$$y'' = \frac{3 \cos t (1 + \sin^2 t)^4}{a \sin^3 t (\sin^2 t - 3)^3}$$

$$r = \frac{a \cos t}{\sqrt{1 + \sin^2 t}}$$

$$\tan \theta = \sin t$$

$$m = \frac{1 - 3 \sin^2 t}{\sin t (\sin^2 t - 3)}$$

$$\tan \psi = -\frac{\cos^2 t}{2 \sin t}$$

$$p = \frac{-a \cos^3 t}{(1 + \sin^2 t)^{3/2}}$$

5.4. Eight Curve

The *eight curve* (Figure 38), also called the *lemniscate of Gerono*, is defined by the equation

5.4.1) ... $$x^4 = a^2 (x^2 - y^2).$$

Its polar equation is easily found to be

5.4.2) ... $$r^2 = a^2 \sec^4\theta \cos 2\theta.$$

The parametric equations are found by letting $y = x \sin t$ in 5.4.1,:

5.4.3) ... $$\begin{cases} x = a \cos t \\ y = a \sin t \cos t \end{cases} \quad -\pi \leq t \leq \pi.$$

Geometry of the Eight Curve.

Intercepts $(0, a, 0)$, $(\pm \frac{\pi}{2}, 0, 0)$, $(\pm \pi, -a, 0)$

Extrema $(0, a, 0)$ is x-maximum

$(\pm \pi, -a, 0)$ is x-minimum

$\left(\frac{\pi}{4}, \frac{a\sqrt{2}}{2}, \frac{a}{2}\right)$, $\left(-\frac{3\pi}{4}, \frac{-a\sqrt{2}}{2}, \frac{a}{2}\right)$ are y-maxima

$\left(-\frac{\pi}{4}, \frac{a\sqrt{2}}{2}, \frac{-a}{2}\right)$, $\left(\frac{3\pi}{4}, \frac{-a\sqrt{2}}{2}, \frac{-a}{2}\right)$ are y-minima

Inflection $(\pm \frac{\pi}{2}, 0, 0)$

Extent $-\pi \leq t \leq \pi$

$-a \leq x \leq a$

$-\frac{a}{2} \leq y \leq \frac{a}{2}$

Symmetries	$y = 0$; $x = 0$; $(0, 0)$
Loops	two: 1) $-\pi \leq t \leq -\frac{\pi}{2}$ and $\frac{\pi}{2} \leq t \leq \pi$
	2) $-\frac{\pi}{2} \leq t \leq \frac{\pi}{2}$
Node	$(\pm \frac{\pi}{2}, 0, 0)$

Analysis of the Eight Curve.

$x = a \cos t$

$y = a \sin t \cos t$

$\dot{x} = -a \sin t$

$\ddot{x} = -a \cos t$

$\dot{y} = a \cos 2t = a(1 - 2\sin^2 t)$

$\ddot{y} = -2a \sin 2t = -4a \sin t \cos t$

$y'' = \dfrac{\cos t (1 + 2\sin^2 t)}{a \sin^3 t}$

$r = a \cos t \sqrt{1 + \sin^2 t}$

$\tan \theta = \sin t$

$m = -\dfrac{1 - 2\sin^2 t}{\sin t}$

$\tan \psi = -\dfrac{1}{2} \dfrac{\cos^2 t}{\sin^3 t}$

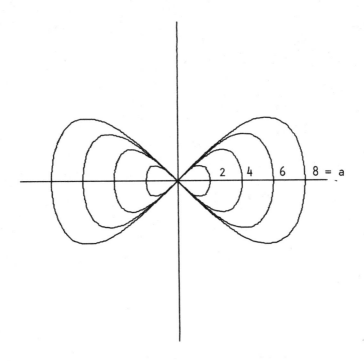

Figure 38. Eight Curve
a = 2, 4, 6, 8

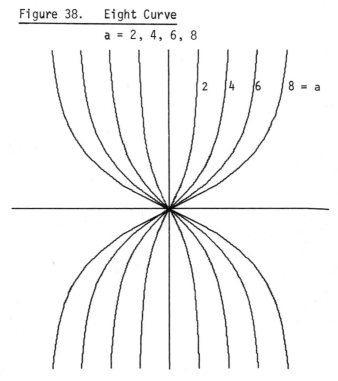

Figure 39. Bullet Nose -1
a = 2, 4, 6, 8;   b = 4

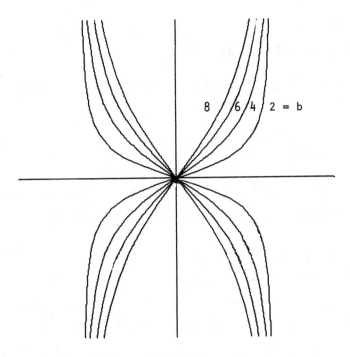

Figure 40. Bullet Nose -2
 a = 6;   b = 2, 4, 6, 8

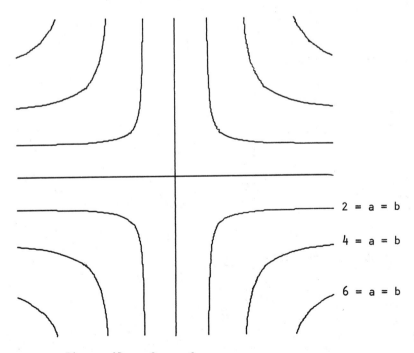

Figure 41. Cross Curve
 a = b = 2, 4, 6

5.5. Bullet Nose (Schoute, 1885)

Through the points of intersection of a tangent to the hyperbola

$$\frac{x^2}{a^2} - \frac{y^2}{b^2} = 1$$

with the x and y axes, lines parallel to each of the axes are drawn. The locus of the intersection of these parallels is the *bullet nose* (Figures 39-40).

Tangents to a hyperbola with slope m are given by $y = mx \pm \sqrt{a^2 m^2 - b^2}$. Hence, the bullet nose is described by

5.5.1) ... $\quad x = \pm \dfrac{\sqrt{a^2 m^2 - b^2}}{m} \quad$ and $\quad y = \pm \sqrt{a^2 m^2 - b^2}$

so the Cartesian equation is

5.5.2) ... $\quad \dfrac{a^2}{x^2} - \dfrac{b^2}{y^2} = 1$.

It has parametric equations

5.5.3) ... $\quad \begin{cases} x = a \cos t \\ y = b \cot t \end{cases} \quad -\pi < t < \pi$

and polar equation

5.5.4) ... $\quad r^2 \sin^2\theta \cos^2\theta = a^2 \sin^2\theta - b^2 \cos^2\theta$.

Geometry of the Bullet Nose.

Intercepts $\quad (\pm \frac{\pi}{2}, 0, 0)$

Inflection $\quad (\pm \frac{\pi}{2}, 0, 0)$

Extent	$-\pi < t < \pi$ $(t \neq 0)$
	$-a < x < a$
	$-\infty < y < \infty$
Discontinuity	$t = 0$
Symmetries	$x = 0$; $y = 0$; $(0, 0)$
Asymptotes	$x = a$
	$x = -a$
Node	$(\pm \frac{\pi}{2}, 0, 0)$

Analysis of the Bullet Nose.

$x = a \cos t$

$y = b \cot t$

$\dot{x} = -a \sin t$

$\ddot{x} = -a \cos t$

$\dot{y} = -b \csc^2 t$

$\ddot{y} = 2b \cos t \csc^3 t$

$y'' = \dfrac{3b}{a^2} \cos t \csc^5 t$

$r = \cos t \sqrt{a^2 + b^2 \csc^2 t}$

$\tan \theta = \dfrac{b}{a} \csc t$

$m = \dfrac{b}{a} \csc^3 t$

5.6. Cross Curve

Through the points of intersection of a tangent to the ellipse

$$\frac{x^2}{a^2} + \frac{y^2}{b^2} = 1$$

with the x- and y-axes, lines parallel to each of the axes are drawn. The locus of the intersection of these parallels is the *cross curve* (Figure 41). Note how similar this definition is to that of the bullet nose.

The Cartesian equation is

5.6.1) ... $$\frac{a^2}{x^2} + \frac{b^2}{y^2} = 1$$

and the parametric equations are

5.6.2) ... $$\begin{cases} x = a \sec t \\ y = b \csc t \end{cases} \quad -\pi < t < \pi .$$

The polar equation is given by

5.6.3) ... $$r^2 \sin^2\theta \cos^2\theta = a^2 \sin^2\theta + b^2 \cos^2\theta .$$

Geometry of the Cross Curve.

Intercepts	None
Extent	$-\pi < t < \pi \quad (t \neq \pm \frac{\pi}{2}, 0)$
	$x > a \quad \text{or} \quad x < -a$
	$y > b \quad \text{or} \quad y < -b$
Discontinuities	$(-\frac{\pi}{2}, \pm\infty, -b)$
	$(0, a, \pm\infty)$
	$(\frac{\pi}{2}, \pm\infty, b)$

Symmetries $x = 0$; $y = 0$ $(0, 0)$

Asymptotes $x = a$; $x = -a$

$y = b$; $y = -b$

Analysis of the Cross Curve.

$x = a \sec t$

$y = b \csc t$

$\dot{x} = a \sec t \tan t$

$\ddot{x} = a \sec^3 t (1 + \sin^2 t)$

$\dot{y} = -b \csc t \cot t$

$\ddot{y} = b \csc^3 t (1 + \cos^2 t)$

$y'' = \dfrac{3b}{a^2} \cos^4 t \csc^5 t$

$r = \sqrt{a^2 \sec^2 t + b^2 \csc^2 t}$

$\tan \theta = \dfrac{b}{a} \cot t$

$m = -\dfrac{b}{a} \cot^3 t$

5.7. Deltoid (Euler, 1745)

The *deltoid* (or *tricuspid*) is the locus of a point P on the circumference of a circle rolling inside a circle three times as large (Figure 42).

Let the fixed circle C be $x^2 + y^2 = 9a^2$, and suppose $P = (3a, 0)$ for $\theta = 0$. Then, after the moving circle C_2 has rolled through an angle θ (that is, the line $O_1 O_2$ connecting the centers of C_1 and C_2 makes an angle θ with the x-axis), the point P has rolled around O_2

through an angle $2\theta = 3\theta - \theta$ (see sketch). If $O_2 = (O_x, O_y)$, then $P = (O_x, O_y) + (a \cos 2\theta, -a \sin 2\theta)$. But, $(O_x, O_y) = (2a \cos \theta, 2a \sin \theta)$.

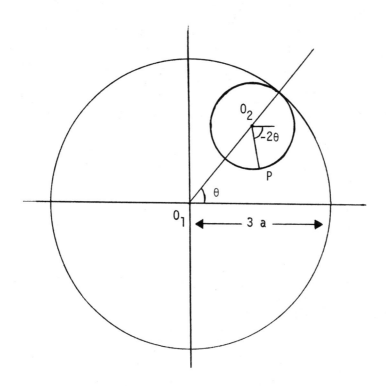

Hence, the parametric equations of the deltoid are

5.7.1) ... $\begin{cases} x = a(2\cos t + \cos 2t) \\ y = a(2\sin t - \sin 2t) \end{cases}$ $-\pi \leq t \leq \pi$.

The Cartesian equation is complicated:

5.7.2) ... $(x^2 + y^2)^2 - 8ax(x^2 - 3y^2) + 18a^2(x^2 + y^2) = 27a^4$

The length of the curve is $L = 16a$, and its area is $A = 2\pi a^2$. The pedal equation is

5.7.3) ... $r^2 + 8 p^2 = 9 a^2$

and the Cesáro form is

5.7.4) ... $9 s^2 + \rho^2 = 64 a^2$.

Whewell's equation is

$$3 s = 8 a \cos 3 \phi .$$

Other results are

$$\rho = - 8 p$$

and

$$p = a \sin 3 \phi .$$

Geometry of the Deltoid.

Intercepts	$(0, 3 a, 0)$, $(\pm \pi, - a, 0)$
	$\left(\pm \cos^{-1} \frac{1}{2} (\sqrt{3} - 1) , \quad 0 , \quad \pm a \sqrt{6\sqrt{3} - 9} \right)$
Extent	$- \pi \leq t \leq \pi$
	$- \frac{3}{2} a \leq x \leq 3 a$
	$- \frac{3\sqrt{3}}{2} a \leq y \leq \frac{3\sqrt{3}}{2} a$
Symmetry	$y = 0$; $y = \pm \sqrt{3} x$
Cusps	$(0, 3 a, 0)$, $\left(\pm \frac{2\pi}{3} , - \frac{3}{2} a, \pm \frac{3\sqrt{3}}{2} a \right)$

Analysis of the Deltoid.

$x = 2a \cos t (1 + \cos t) - a$

$y = 2a \sin t (1 - \cos t)$

$\dot{x} = -2a \sin t (1 + 2\cos t)$

$\ddot{x} = 4a - 2a \cos t (1 + 4\cos t)$

$\dot{y} = 2a + 2a \cos t (1 - 2\cos t) = 2a (1 + 2\cos t)(1 - \cos t)$

$\ddot{y} = 2a \sin t (4 \cos t - 1)$

$y'' = \dfrac{1 - \cos t}{2a \sin^3 t (1 + 2 \cos t)}$

$r = a \sqrt{16 \cos^3 t - 12 \cos t + 5}$

$\tan \theta = \dfrac{2 \sin t (1 - \cos t)}{2 \cos t (1 + \cos t) - 1}$

$m = \dfrac{\cos t - 1}{\sin t} = \tan (\pi - \tfrac{1}{2} t)$

$\tan \psi = \dfrac{(1 + 2 \cos t)(1 - \cos t)}{3 \sin t (1 - 2 \cos t)}$

$p = \tfrac{1}{2} a (1 + 2 \cos t) \sqrt{2(1 - \cos t)}$

$s = \tfrac{8}{3} a \cos \tfrac{3}{2} t = \tfrac{4}{3} a \sqrt{2} (1 - 2 \cos t) \sqrt{1 + \cos t}$

$\rho = -8a \sin \tfrac{3}{2} t$

$\alpha = 3a (1 + 2 \cos t - 2 \cos^2 t)$

$\beta = 6a \sin t (1 + \cos t)$

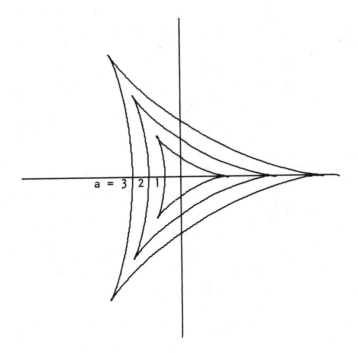

Figure 42.   Deltoid
a = 1, 2, 3

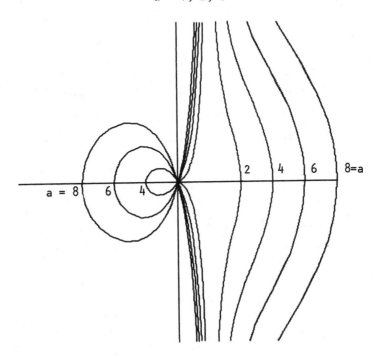

Figure 43.   Conchoid of Nicomedes -1
a = 2, 4, 6, 8;   b = 2

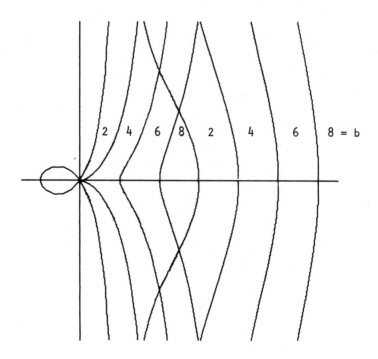

Figure 44. Conchoid of Nicomedes -2
a = 4;   b = 2, 4, 6, 8

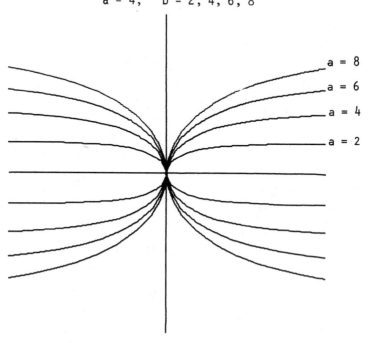

Figure 45. Kappa Curve
a = 2, 4, 6, 8

5.8. Conchoid of Nicomedes (Nicomedes, 225 B.C.)

The *conchoid of Nicomedes* (also known as the *cochloid*) is the conchoid of a line with respect to a point not on the line (Figures 43-44). If the line is $x = b$ and the point is $(0, 0)$, then the conchoid, with parameter a, is (by 2.5.8)

$$x = b \pm \frac{a b}{\sqrt{t^2 + b^2}}$$

$$y = t \pm \frac{a t}{\sqrt{t^2 + b^2}} \ .$$

Substituting $\hat{t} = b \tan t$, we have the parametric equations

5.8.1) ... $\qquad \begin{cases} x = b + a \cos t \\ y = \tan t \ (b + a \cos t) \end{cases} \qquad -\frac{\pi}{2} < t < \frac{3\pi}{2} \ .$

Now, $r^2 = x^2 + y^2$; carrying out the algebra results in the polar equation

5.8.2) ... $\qquad r = a + b \sec t \ .$

This may also be found by representing the line in polar coordinates 1.1.8 as $r \cos \theta = b$ and using 2.5.10.

Combining the last two results, allows the Cartesian equation to be obtained:

5.8.3) ... $\qquad (x^2 + y^2)(x - b)^2 = a^2 x^2 \ .$

If $b < a$, there is a loop with area

$$A = b\sqrt{a^2 - b^2} - 2 a b \ \ln\left(\frac{a + \sqrt{a^2 - b^2}}{b}\right) + a^2 \cos^{-1} \frac{b}{a} \ .$$

Geometry of the Conchoid.

Intercepts	$(0, a+b, 0)$, $(\pi, b-a, 0)$, $(\pm\pi \mp \cos^{-1}\frac{b}{a}, 0, 0)$
Extrema	$(0, a+b, 0)$ is x-maximum
	$(\pm\pi, b-a, 0)$ is x-minimum
	$\cos t = \sqrt{-\frac{b}{a}}$ defines y-extrema (if $b < a$)
Extent	$-\pi \leq t \leq \pi$
	$b - a \leq x \leq a + b$
	$-\infty < y < \infty$
Symmetry	$y = 0$
Asymptote	$x = a$
Loop	if $b < a$, $b - a \leq x \leq 0$ is a loop
Isolated Point	$(0, 0)$ in 5.8.3 if $b > a$
Node	$(0, 0)$ if $b < a$
Cusp (1st kind)	$(0, 0)$ if $b = a$

Analysis of the Conchoid.

$x = b + a \cos t$

$y = \tan t \, (b + a \cos t)$

$\dot{x} = -a \sin t$

$\ddot{x} = -a \cos t$

$\dot{y} = b \sec^2 t + a \cos t$

$\ddot{y} = 2 b \sin t \sec^3 t - a \sin t$

$y'' = \dfrac{2b - 3b\cos^2 t - a\cos^3 t}{a^2 \sin^3 t \cos^3 t}$

$$r = \frac{b + a \cos t}{\cos t}$$

$$\tan \theta = \tan t$$

$$m = -\frac{b + a \cos^3 t}{a \sin t \cos^2 t}$$

$$\tan \psi = \frac{\cos t \, (b + a \cos t)}{b \sin t}$$

5.9. Kappa Curve

The *kappa curve* (or *Gutschoven's curve*) is the locus of points P found as follows. Let $\mathcal{L}$ be the lines $y = a$, intersecting the y-axis at C. From $O = (0, 0)$, draw an arbitrary line $\hat{\mathcal{L}}$ meeting $\mathcal{L}$ at D. P is the locus of points on $\hat{\mathcal{L}}$ such that $OP = CD$. (See Figure 45.)

If $\hat{\mathcal{L}}$ is given by $x = m y$; then $D = (a m, a)$, so $\overline{CD} = a m$. But, $\overline{OP} = \sqrt{x^2 + y^2} = \overline{CD}$, so

$$x^2 + y^2 = a^2 m^2 = a^2 x^2/y^2 .$$

Thus, the Cartesian equation is

5.9.1) ... $\qquad (x^2 + y^2) y^2 = a^2 x^2 .$

The polar equation is clearly

5.9.2) ... $\qquad r = a \cot \theta ,$

and the parametric equations are

5.9.3) ... $\qquad \begin{cases} x = a \cos t \cot t \\ y = a \cos t \end{cases} \quad 0 < t < 2\pi .$

Geometry of the Kappa Curve.

Intercepts	$(\frac{\pi}{2}, 0, 0)$, $(\frac{3\pi}{2}, 0, 0)$
Extrema	$(\frac{\pi}{2}, 0, 0)$ is x-minimum
	$(\frac{3}{2}\pi, 0, 0)$ is x-maximum
Extent	$0 < t < 2\pi$
	$-\infty < x < \infty$
	$-a < y < a$
Symmetry	$x = 0$; $y = 0$; $(0, 0)$
Asymptotes	$y = a$; $y = -a$
Double Cusp	$(0, 0)$

Analysis of the Kappa Curve.

$x = a \cos t \cot t$

$y = a \cos t$

$\dot{x} = -a \cos t (1 + \csc^2 t)$

$\ddot{x} = a (2 \csc^3 t - \csc t + \sin t)$

$\dot{y} = -a \sin t$

$\ddot{y} = -a \cos t$

$y'' = \dfrac{(\sin^2 t - 3) \sin^4 t}{a \cos^3 t (\sin^2 t + 1)^3}$

$r = a \cot t$

$\tan \theta = \tan t$

$m = \dfrac{\sin^3 t}{\cos t (\sin^2 t + 1)}$

$\tan \psi = -\sin t \cos t$

$$p = \frac{-a\cos^2 t}{\sqrt{1 + \sin^2 t - \sin^4 t}}$$

5.10. Kampyle of Eudoxus

The *kampyle of Eudoxus* is defined to be the locus of

5.10.1) ... $\quad x^4 = a^2 (x^2 + y^2)$.

Its polar equation is clearly

5.10.2) ... $\quad r \cos^2\theta = a$,

and the parametric equations, with $t = \theta$, are

5.10.3) ... $\quad \begin{cases} x = a \sec t \\ y = a \tan t \sec t \end{cases} \quad -\frac{\pi}{2} < t < \frac{3\pi}{2}$.

It is illustrated in Figure 46.

Geometry of the Kampyle of Eudoxus.

Intercepts	$(0, a, 0)$, $(\pi, -a, 0)$
Extrema	$(0, a, 0)$ is x-minimum
	$(\pi, -a, 0)$ is x-maximum
Extent	$-\pi \leq t \leq \pi$
	$x \leq -a$; $x \geq a$
	$-\infty < y < \infty$
Inflection	$\left(\tan^{-1} \pm \frac{\sqrt{2}}{2}, \pm a \frac{\sqrt{6}}{2}, \pm a \frac{\sqrt{3}}{2}\right)$ [4 points of inflection]

Symmetry $x = 0$; $y = 0$; $(0, 0)$

Isolated Point $(0, 0)$ in 5.10.1.

Analysis of the Kampyle of Eudoxus.

$x = a \sec t$

$y = a \sec t \tan t$

$\dot{x} = a \sec t \tan t$

$\ddot{x} = a \sec t (1 + 2 \tan^2 t)$

$\dot{y} = a \sec t (1 + 2 \tan^2 t)$

$\ddot{y} = a \sec t \tan t (5 + 6 \tan^2 t)$

$y'' = \dfrac{1}{a} \sec t \cot^3 t (2 \tan^2 t - 1)$

$r = a \sec^2 t$

$\theta = t$

$m = \dfrac{1 + \sin^2 t}{\sin t \cos t}$

$\tan \psi = \dfrac{1}{2} \cot t$

$p = \dfrac{-a}{\cos t \sqrt{3 \sin^2 t + 1}}$

$\rho = \dfrac{a (3 \sin^2 t + 1)^{3/2}}{\cos^3 t (3 \sin^2 t - 1)}$

$\alpha = \dfrac{-2a (1 + 3 \sin^4 t)}{\cos^3 t (3 \sin^2 t - 1)}$

$\beta = \dfrac{6 a \sin^3 t}{\cos^2 t (3 \sin^2 t - 1)}$

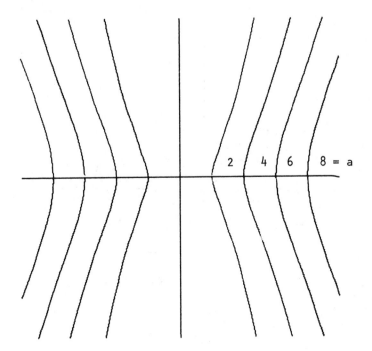

Figure 46. Kampyle of Eudoxus
a = 2, 4, 6, 8

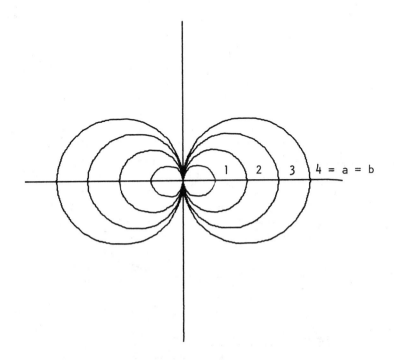

Figure 47. Hippopede -1
a = b = 1, 2, 3, 4

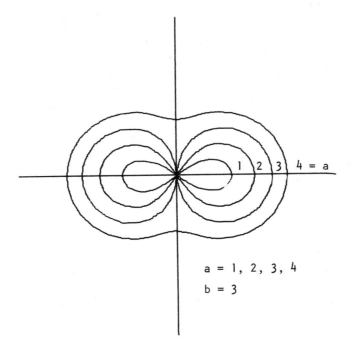

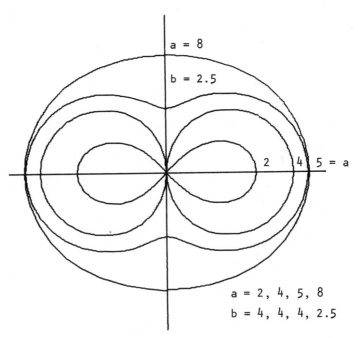

Figure 48. Hippopede -2
a = 1, 2, 3, 4; b = 3
a = 2, 4, 5, 8; b = 4, 4, 4, 2.5

5.11. Hippopede (Proclus, ca. 75 B.C.)

The *hippopede*, also known as the *horse fetter*, is the locus of

$$5.11.1) \quad (x^2 + y^2)^2 + 4b(b-a)(x^2 + y^2) = 4b^2 x^2,$$

where $a > 0$, $b > 0$. The polar equation is clearly

$$5.11.2) \quad r^2 = 4b(a - b\sin^2\theta).$$

If $b = 2a$, the curve reduces to the lemniscate of Bernoulli. (See Figures 47-48.)

The parametric equations, with parameter $\theta = t$, are easily shown to be

$$5.11.3) \quad \begin{cases} x = 2\cos t \sqrt{ab - b^2 \sin^2 t} \\ y = 2\sin t \sqrt{ab - b^2 \sin^2 t} \end{cases} \quad -\pi \leq t \leq \pi.$$

However, this form loses the isolated point at $(0, 0)$ if $a > b$. If $a \leq b$, the curve forms a figure eight. If $b < a \leq 2b$, it is an oval with indentations at top and bottom, and if $2b \leq a$, it is an oval.

If $a < b$, the curve does not exist for all values of t between $-\pi$ and π. The right-hand loop corresponds to $-\arcsin\sqrt{\frac{a}{b}} \leq t \leq \arcsin\sqrt{\frac{a}{b}}$, and the left-hand loop to $\pi - \sin^{-1}\sqrt{\frac{a}{b}} \leq t \leq \pi$ and $-\pi \leq t \leq -\pi + \sin^{-1}\sqrt{\frac{a}{b}}$.

Geometry of the Hippopede.

Intercepts $\quad (0, 2\sqrt{ab}, 0)$, $\left(\pm\frac{\pi}{2}, 0, 2\sqrt{ab - b^2}\right)^*$,
$\quad\quad\quad\quad\quad (\pm\pi, -2\sqrt{ab}, 0)$

Extrema $\quad\quad (0, 2\sqrt{ab}, 0)$ is x-maximum

---

*$a \geq b$.

$(\pm \pi, -2\sqrt{ab}, 0)$ is x-minimum

y-extrema are for $a = 2b \sin^2 t$ $\qquad (a \leq 2b)$

and $\left(\pm \frac{\pi}{2}, 0, \pm 2\sqrt{ab - b^2}\right)$ $\qquad (a > b)$

At the first point, $x = \pm\sqrt{2b - a}$ and $y = \pm a$

Symmetric about $\quad x = 0, \quad y = 0, \quad (0, 0)$

Singularities: $\quad a < b$, Node at $(0, 0)$

$\qquad\qquad\quad a = b$, Double Cusp at $(0, 0)$

$\qquad\qquad\quad a > b$, Isolated Point at $(0, 0)$

Analysis of the Hippopede.

$x = 2 \cos t \sqrt{ab - b^2 \sin^2 t}$

$y = 2 \sin t \sqrt{ab - b^2 \sin^2 t}$

$\dot{x} = \dfrac{2b(a - b) \sin t}{\sqrt{ab - b^2 \sin^2 t}}$

$\dot{y} = \dfrac{2b(a - 2b \sin^2 t) \cos t}{\sqrt{ab - b^2 \sin^2 t}}$

$\ddot{x} = \dfrac{2ab^2(a - b) \cos t}{(ab - b^2 \sin^2 t)^{3/2}}$

$\ddot{y} = \dfrac{2b^2 \sin t [4b^2 \cos^4 t + (5ab - 2b^2) \sin^2 t - a^2 - 2ab]}{(ab - b^2 \sin^2 t)^{3/2}}$

5.12. Bicorn (Sylvester, 1864)

The *bicorn* (Figure 49) is defined as the locus of the equation

5.12.1) ... $\quad (x^2 + 2ay - a^2)^2 = y^2(a^2 - x^2)$.

Now, the right-hand side of this equation must be nonnegative, or $|x| \leq |a|$. Hence, let $x = a \sin t$. Then, 5.12.1 yields the parametric equations

5.12.2) ... $\quad \begin{cases} x = a \sin t \\ y = \dfrac{a \cos^2 t\ (2 + \cos t)}{3 + \sin^2 t} \end{cases} \quad -\pi \leq t \leq \pi$.

Geometry of the Bicorn.

Intercepts $\quad (0, a, 0)$, $(\frac{\pi}{2}, 0, a)$, $(\pm \pi, -a, 0)$,

$(-\frac{\pi}{2}, 0, \frac{1}{3}a)$

Extrema $\quad (\frac{\pi}{2}, 0, a)$, $(-\frac{\pi}{2}, 0, \frac{1}{3}a)$ $\quad$ y-maxima

Extent $\quad -\pi \leq t \leq \pi$
$\quad\quad\quad\quad -a \leq x \leq a$
$\quad\quad\quad\quad 0 \leq y \leq a$

Cusps $\quad (\pm a, 0)$

Symmetry $\quad x = 0$

Analysis of the Bicorn.

$x \quad = a \sin t$

$y \quad = a \cos^2 t\ \dfrac{2 + \cos t}{3 + \sin^2 t}$

$\dot{x} \quad = a \cos t$

$\ddot{x} \quad = -a \sin t$

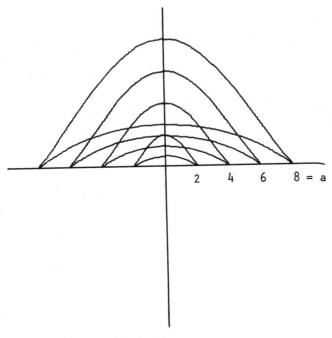

Figure 49. Bicorn
a = 2, 4, 6, 8

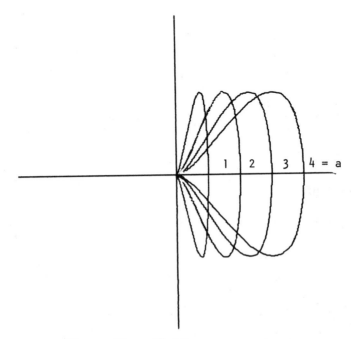

Figure 50. Piriform
a = 1, 2, 3, 4; b = 4

$$\dot{y} = -a \sin t \cos t \frac{16 + 12 \cos t - \cos^3 t}{(3 + \sin^2 t)^2}$$

$$\ddot{y} = a \frac{64 + 96 \cos t - 80 \cos^2 t - 136 \cos^3 t - 32 \cos^4 t + 8 \cos^5 t - \cos^7 t}{(3 + \sin^2 t)^3}$$

5.13. Piriform (De Longchamps, 1886)

The *piriform*, also known as the *pear-shaped quartic* (Figure 50), is defined to be the locus of a point P satisfying the following definition. Let C be a circle and $\mathcal{L}$ a line. Let O be on the circumference of C such that the diameter through O is perpendicular to $\mathcal{L}$. Draw an arbitrary line $\hat{\mathcal{L}}$ through O, intersecting $\mathcal{L}$ at P_1. Draw a line $\mathcal{L}_1$ perpendicular to $\mathcal{L}$ through P_1, intersecting C at P_2. Draw a line $\mathcal{L}_2$ parallel to $\mathcal{L}$ through P_2; the intersection of $\mathcal{L}_2$ and the arbitrary line $\hat{\mathcal{L}}$ is the point P.

Let the circle be given by $(x - a)^2 + y^2 = a^2$, $O = (0, 0)$, and $\mathcal{L}$ be given by $x = a^2/b$. Then, if $\hat{\mathcal{L}}$ is $y = m x$,

$$P_1 = \left(\frac{a^2}{b}, m \frac{a^2}{b}\right) \quad \text{and} \quad P_2 = \left(a \pm \sqrt{a^2 - m^2 \frac{a^4}{b^2}}, m \frac{a^2}{b}\right).$$

Thus,

$$P = \left(a \pm \frac{a}{b}\sqrt{b^2 - m^2 a^2}, m a \pm m \frac{a}{b}\sqrt{b^2 - m^2 a^2}\right).$$

Now, if we define t by $m = \frac{b}{a} \cos t$, $-\frac{\pi}{2} \leq t \leq \frac{\pi}{2}$, we have the parametric equations of the piriform,

5.13.1) ... $\quad \begin{cases} x = a (1 + \sin t) \\ y = b \cos t (1 + \sin t) \end{cases} \quad -\frac{\pi}{2} \leq t \leq \frac{3\pi}{2}$.

By eliminating t from 5.13.1, the Cartesian equation

5.13.2) ... $\quad a^4 y^2 = b^2 x^3 (2a - x)$

results. The polar equation is

5.13.3) ... $\quad b^2 r^2 \cos^4\theta - 2ab^2 r \cos^3\theta + a^4 \sin^2\theta = 0$.

Geometry of the Piriform.

Intercepts $\quad (-\frac{\pi}{2}, 0, 0)$, $(\frac{\pi}{2}, 2a, 0)$, $(\frac{3\pi}{2}, 0, 0)$

Extent $\quad -\frac{\pi}{2} \leq t \leq \frac{3\pi}{2}$

$0 \leq x \leq 2a$

$-\frac{3b\sqrt{3}}{4} \leq y \leq \frac{3b\sqrt{3}}{4}$

Symmetry $\quad y = 0$

Cusp $\quad (0, 0)$

Analysis of the Piriform.

$x \quad = a(1 + \sin t)$

$y \quad = b \cos t (1 + \sin t)$

$\dot{x} \quad = a \cos t$

$\ddot{x} \quad = -a \sin t$

$\dot{y} \quad = b(1 + \sin t)(1 - 2 \sin t)$

$\ddot{y} \quad = -b \cos t (1 + 4 \sin t)$

$y'' \quad = \dfrac{b(2 \sin^3 t - 3 \sin t - 1)}{a^2 \cos^3 t}$

$r \quad = (1 + \sin t)\sqrt{a^2 + b^2 \cos^2 t}$

$\tan \theta = \dfrac{b}{a} \cos t$

$m \quad = \dfrac{b}{a} \sec t (1 + \sin t)(1 - 2 \sin t)$

5.14. Devil's Curve (Cramer, 1750)

The *Devil's curve* (or the *Devil on two sticks*) is the locus of

5.14.1) ... $\quad y^4 - a^2 y^2 = x^4 - b^2 x^2 .$

The polar equation is

5.14.2) ... $\quad r^2 (\sin^2\theta - \cos^2\theta) = a^2 \sin^2\theta - b^2 \cos^2\theta ,$

and the parametric equations are

5.14.3) ... $\begin{cases} x = \cos t \sqrt{\dfrac{a^2 \sin^2 t - b^2 \cos^2 t}{\sin^2 t - \cos^2 t}} \\ \\ y = \sin t \sqrt{\dfrac{a^2 \sin^2 t - b^2 \cos^2 t}{\sin^2 t - \cos^2 t}} \end{cases}$

$\quad -\dfrac{\pi}{4} < t < \dfrac{\pi}{4}$

$\quad \dfrac{3}{4}\pi < t < \dfrac{5}{4}\pi$

$\quad u \leq t \leq \pi - u$

$\quad -(\pi - u) \leq t \leq -u$

where $u = \tan^{-1} \dfrac{b}{a} .$

See Figure 51.

Geometry of the Devil's Curve.

Intercepts	$(0, \pm a)$, $(\pm b, 0)$, $(0, 0)$
Symmetry	$x = 0$, $y = 0$
Asymptote	$x \pm y = 0$
Node	$(0, 0)$

5.15. Folia (Kepler, 1609)

The *folia* are defined as pedals of the deltoid. The general equation is

5.15.1) ... $\quad [(x - b)^2 + y^2] [x(x - b) + y^2] = 4 a (x - b) y^2 .$

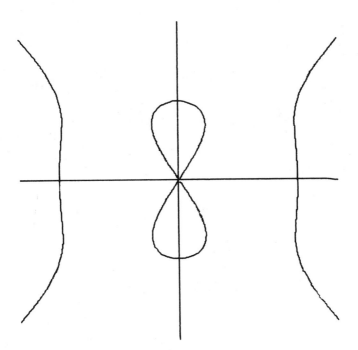

Figure 51. Devil's Curve
a = 2; b = 3

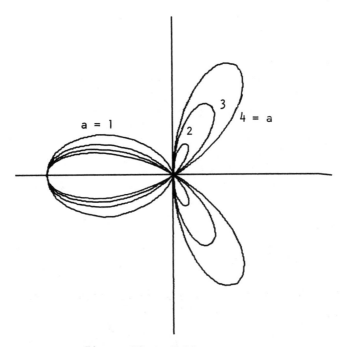

Figure 52. Folium
a = 1, 2, 3, 4; b = 4

Translating the origin to (b, 0) results in the polar equation

5.15.2) ... $\quad r = \cos \theta \, (4 a \sin^2 \theta - b)$.

If $b \geq 4a$, the folium is single. If $b = 0$, it is double. If $0 < b < 4a$, it is a trifolium. Illustrations are given in Figure 52.

Geometry of the Folia.

Intercepts	(0, 0) , (-b, 0)
Symmetry	$y = 0$
Node	(0, 0)

5.16. Cassinian Oval (Cassini, 1680)

The *Cassinian oval* (also termed the *Cassinian ellipse*) is the locus of a point P whose bipolar coordinates r_1 and r_2 (with respect to foci F_1 and F_2 a distance 2 a apart) satisfy the relation

5.16.1) ... $\quad r_1 r_2 = b^2$.

There are, thus, two parameters (a and b).

If $a = b$, the oval is a Lemniscate of Bernoulli. If $a < b$, there is one loop and if $a \geq b$, two loops.

The Cartesian equation is easily derived from equation 1.3.7, yielding

$$[(x - a)^2 + y^2] \, [(x + a)^2 + y^2] = b^4$$

$$x^4 + y^4 + a^4 + 2 \, (x^2 y^2 + a^2 y^2 - a^2 x^2) = b^4$$

5.16.2) ... $\quad (x^2 + y^2 + a^2)^2 = b^4 + 4 a^2 x^2$.

a = 5.5

a = 6

a = 6.5

Figure 53. Cassinian Oval
a = 5.5, 6, 6.5
b = 6

The polar equation is

5.16.3) ... $\quad r^4 - 2a^2 r^2 \cos 2\theta = b^4 - a^4$.

The curve is illustrated in Figure 53.

Geometry of the Cassinian Oval.

Intercepts $\quad \left(\pm\sqrt{a^2 \pm b^2},\, 0\right),\; \left(0,\, \pm\sqrt{b^2 - a^2}\right)$

Extrema $\quad \left(\pm\sqrt{a^2 \pm b^2},\, 0\right),\; \left(0,\, \pm\sqrt{b^2 - a^2}\right)^*,$

$\qquad\qquad \left(\pm\dfrac{\sqrt{4a^4 - b^4}}{2a},\; \pm\dfrac{b^2}{2a}\right)$

Symmetries $\quad x = 0;\quad y = 0;\quad (0, 0)$

Loops $\qquad$ one if $a < b$
$\qquad\qquad$ two if $a \geq b$

Node $\qquad$ $(0, 0)$ if $a = b$

5.17. Cartesian Oval (Descartes, 1637)

The *Cartesian oval* is the locus of a point P whose bipolar coordinates r_1 and r_2 (with respect to foci F_1 and F_2 a distance 2a apart) satisfy the relation

5.17.1) ... $\quad m\, r_1 + n\, r_2 = 1$.

If $m = n$, the oval becomes an ellipse. If $m = -n$, the oval is a hyperbola. If $a n = 1$, the oval is a Limacon of Pascal. (See Figure 54.)

$^*b \neq a$

-156-

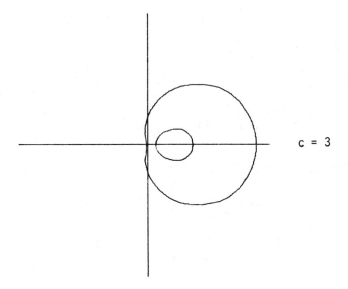

c = 3

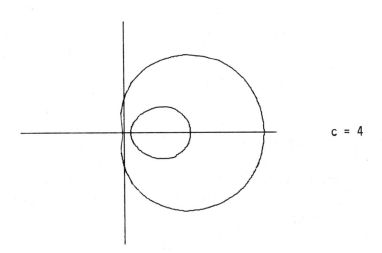

c = 4

Figure 54. Cartesian Oval
b = 1
c = 3, 4

The Cartesian equation may be found using equation 1.3.7, with the origin as the midpoint of the line segment $F_1 F_2$:

$$m\sqrt{(x - a)^2 + y^2} + n\sqrt{(x + a)^2 + y^2} = 1.$$

Transposing the second term to the right-hand side, squaring, and simplifying, gives

$$(x^2 + y^2 + a^2)(m^2 - n^2) - 2ax(m^2 + n^2) - 1$$
$$= -2n\sqrt{(x + a)^2 + y^2}.$$

Squaring and simplifying gives the Cartesian equation:

5.17.2) ... $$[(m^2 - n^2)(x^2 + y^2 + a^2) - 2ax(m^2 + n^2)]^2$$
$$= 2(m^2 + n^2)(x^2 + y^2 + a^2) - 4ax(m^2 - n^2) - 1.$$

Now, let us set $a = 1$, and define

$$b = m^2 - n^2$$
$$c = m^2 + n^2.$$

(Note that $c > b$ and $c > 0$.) Then, 5.17.2 becomes

5.17.3) ... $$[b(x^2 + y^2 + 1) - 2cx]^2 = 2c(x^2 + y^2 + 1) - 4bx - 1.$$

5.18. Dürer's Conchoid (Dürer, 1525)

Let $Q = (q, 0)$ and $R = (0, r)$ be points such that $q + r = b$. On $\overline{QR}$, extended in both directions, mark points P, P' whose distance from Q is equal to a. The locus of points P and P' is *Dürer's conchoid* (Figure 55).

The equation may be found by eliminating q and r from three equations:

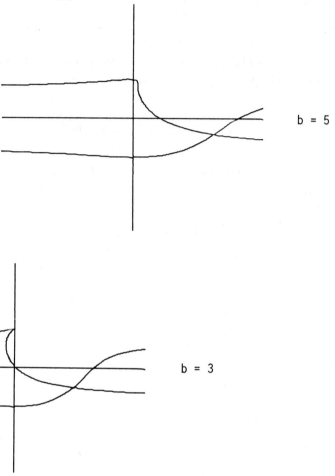

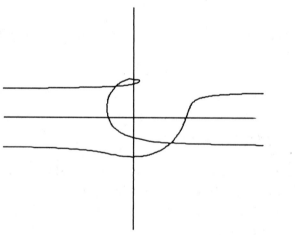

Figure 55. Dürer's Conchoid
a = 3
b = 1, 3, 5

$$b = q + r$$
$$a^2 = (x - q)^2 + y^2$$
$$y = -\frac{r}{q} x + r .$$

It is

5.18.1) ...
$$2 y^2 (x^2 + y^2) - 2 b y^2 (x + y) + (b^2 - 3 a^2) y^2$$
$$- a^2 x^2 + 2 a^2 b (x + y) + a^2 (a^2 - b^2) = 0 .$$

If $a = 0$, the equation reduces to the line $y = 0$. If $b = 0$, it reduces to two parallel lines $y = \pm \sqrt{2}/2\, a$ and a circle $x^2 + y^2 = a^2$.

If $a > b$, there is a loop. If $a = b$, there is a cusp at $(0, a)$. There are two branches asymptotic to $y = \pm \sqrt{2}/2\, a$.

Geometry of Dürer's Conchoid.

Intercepts
  $(b \pm a, 0)$ are x-intercepts
  $(0, \pm a)$ are the extreme y-intercepts
  $\left(0, \frac{1}{2}\left[b \pm \sqrt{2 a^2 - b^2}\right]\right)$ are y-intercepts if $2 a^2 \geq b^2$

Cusp
  $(0, a)$ if $a = b$

Asymptote
  $y = \pm \frac{\sqrt{2}}{2} a$

CHAPTER 6

ALGEBRAIC CURVES OF HIGH DEGREE

The final two chapters discuss and illustrate more complex curves. The sections on geometric and analytic properties have frequently been deleted because of this complexity.

6.1. Epitrochoid (Dürer, 1525)

The word "trochoid" means "roulette"; the *epitrochoid* is the roulette traced by a point P attached to a circle S rolling about the outside of a fixed circle C (Figures 56-58).

Let
  b = radius of S
  a = radius of C
  h = distance from P to the center of S
  m = a + b .

Then the equation of C is $x^2 + y^2 = a^2$, and the equation of S (for t = 0) is $(x - m)^2 + y^2 = b^2$. The value of P (for t = 0) is $P_0 = (m - h, 0)$; see sketch.

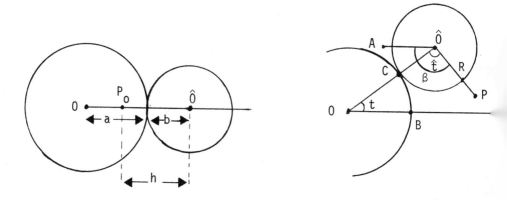

After S rolls around C so that the point of contact is $C = (a \cos t, a \sin t)$, we have

$$P = \hat{O} - h(\cos\beta, \sin\beta)$$

where $\beta = \angle P\hat{O}A$, and

$$\hat{O} = m(\cos t, \sin t).$$

Hence,

$$P = m(\cos t, \sin t) - h(\cos\beta, \sin\beta).$$

Now, $\angle O\hat{O}A = t$, since $A\hat{O}$ is parallel to OB, and arc BC = arc RC. Hence,

$$a\,t = b\,\hat{t},$$

and so,

$$\beta = \hat{t} + \angle O\hat{O}A = \frac{a}{b}t + t = \frac{m}{b}t.$$

Therefore, the parametric equations of the epitrochoid are

6.1.1) ... $\begin{cases} x = m\cos t - h\cos\frac{m}{b}t \\ y = m\sin t - h\sin\frac{m}{b}t \end{cases}$ $-\pi \leq t \leq \pi$.

Special cases are:

Limacon	$a = b$
Circle	$a = 0$
Epicycloid	$h = b$.

There are $\frac{m}{b} - 1$ inner loops if m/b is an integer. The curve is symmetric about the x-axis, and is symmetric about the y-axis if m/b is an odd integer. It is completely contained within a circle defined by $|r| \leq m + h$.

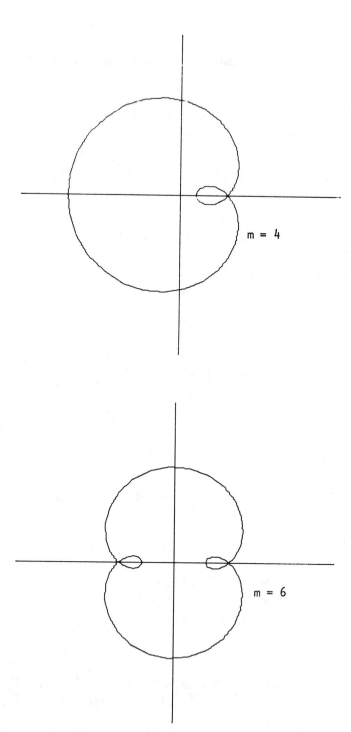

Figure 56. Epitrochoid -1
b = 2; m = 4, 6; h = 3

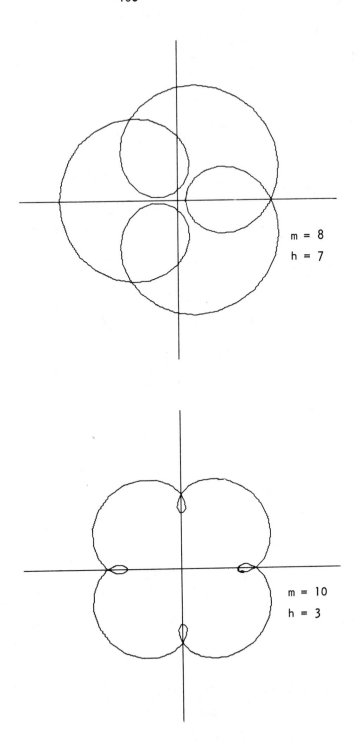

Figure 57. Epitrochoid -2
b = 2; m = 8, 10; h = 7, 3

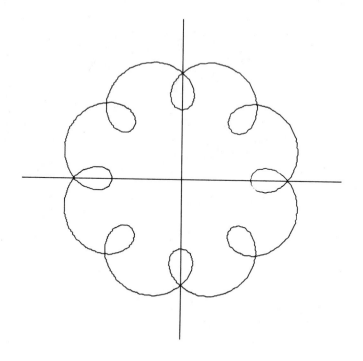

Figure 58. Epitrochoid -3
b = 2; h = 5; m = 18

6.2. Hypotrochoid

A *hypotrochoid* is the roulette traced by a point P attached to a circle S rolling about the inside of a fixed circle C (Figures 59-60).

Let
        b = radius of S
        a = radius of C
        h = distance from P to the center of S
        n = a - b .

Then, the equation of C is $x^2 + y^2 = a^2$, and the equation of S (for t = 0) is $(x - n)^2 + y^2 = b^2$. The value of P (t = 0) is $P_0 = (n - h, 0)$; see sketch.

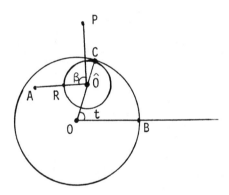

After S rolls around C so that the point of contact is C = (a cos t, a sin t), we have

$$P = \hat{O} + h (\cos \beta, - \sin \beta)$$

where $\beta = \angle A \hat{O} P$, and

$$\hat{O} = n (\cos t, \sin t) .$$

Hence,

$$P = n (\cos t, \sin t) .$$

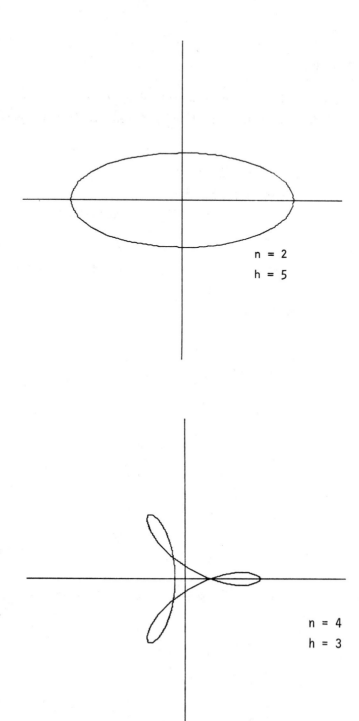

Figure 59. Hypotrochoid -1
b = 2; n = 2, 4; h = 5, 3

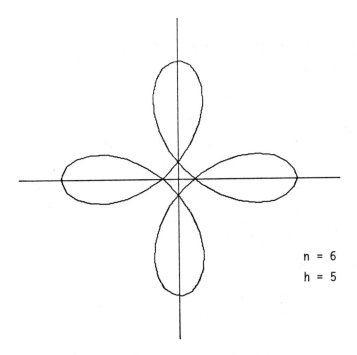

n = 6
h = 5

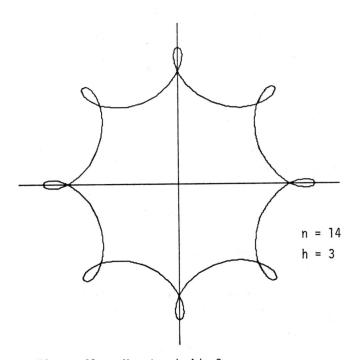

n = 14
h = 3

Figure 60. Hypotrochoid -2
b = 2; n = 6, 14; h = 5, 3

Now, $<O\hat{O}A = t$, and arc BC = arc RC. Hence, $at = b\hat{t}$, where $\hat{t} = <O\hat{O}P$, and so

$$\beta = \hat{t} - <O\hat{O}A = \frac{a}{b}t - t = \frac{n}{b}t.$$

Therefore, the parametric equations of the hypotrochoid are

6.2.1) ... $\begin{cases} x = n \cos t + h \cos \frac{n}{b} t \\ y = n \sin t - h \sin \frac{n}{b} t \end{cases}$ $-\pi \le t \le \pi$.

Special cases are:

Hypocycloid	$h = b$	
Ellipse	$a = 2b$	
Rhodonae	$a = \frac{2nh}{n+1}$,	$b = \frac{(n-1)h}{(n+1)}$.

There are $\frac{n}{b} + 1$ outer loops if n/b is an integer. The curve is symmetric about the y-axis if n/b is an odd integer. It is completely contained within a circle defined by $|r| \le n + h$.

6.3. Epicycloid (Roemer, 1674)

The *epicycloid* (Figure 61) is an epitrochoid with $h = b$; the parametric equations are

6.3.1) ... $\begin{cases} x = m \cos t - b \cos \frac{m}{b} t \\ y = m \sin t - b \sin \frac{m}{b} t \end{cases}$ $-\pi \le t \le \pi$,

where $m = a + b$. If $\frac{a}{b}$ is rational, the curve is algebraic; otherwise, the curve is transcendental. Special cases are:

Cardioid	$a = b$
Nephroid	$a = 2b$.

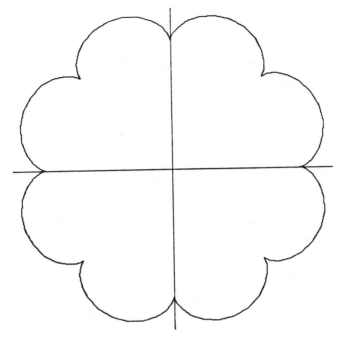

Figure 61. Epicycloid
b = 1; m = 9

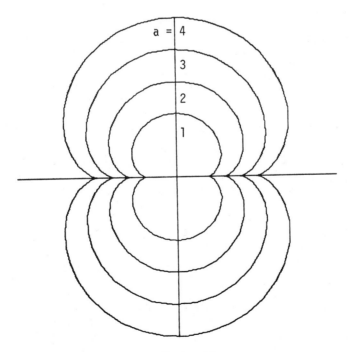

Figure 62. Nephroid
a = 1, 2, 3, 4

There are $\frac{m}{b} - 1$ cusps, if m/b is an integer. The curve is symmetric about the x-axis, and is symmetric about the y-axis if m/b is an odd integer. It is completely contained within a circle defined by $|r| \leq m + b$. Its length is L = 8 m (if m/b is an integer), and its area is A = π m (2 m - b).

6.4. Nephroid (Huygens, 1678)

The *nephroid* (Figure 62) is an epicycloid of two cusps. The parametric equations are

6.4.1) ... $\begin{cases} x = a\ (3 \cos t - \cos 3 t) \\ y = a\ (3 \sin t - \sin 3 t) \end{cases}$ $\quad -\pi \leq t \leq \pi$.

The Cartesian equation is easily seen to be

6.4.2) ... $(x^2 + y^2 - 4 a^2)^3 = 108\ a^4\ y^2$.

The polar equation is

6.4.3) ... $\left(\dfrac{r}{2\,a}\right)^{2/3} = \left(\sin \dfrac{1}{2} \theta\right)^{2/3} + \left(\cos \dfrac{1}{2} \theta\right)^{2/3}$.

Other interesting results include:

6.4.4) ... $\quad 4\ r^2 - 3\ p^2 = 16\ a^2 \quad$ (pedal)

6.4.5) ... $\quad s = 6\ a \sin \dfrac{1}{2} \phi \quad$ (Whewell)

6.4.6) ... $\quad 4\ \rho^2 + s^2 = 36\ a^2 \quad$ (Cesáro)

6.4.7) ... $\quad s = 6\ a\ (1 - \cos \dfrac{1}{2} \psi)\ , \quad 0 < \psi < 2\pi$.

The curve has area A = 12 $\pi\ a^2$ and length L = 24 a.

Geometry of the Nephroid.

Intercepts	$(0, 2a, 0)$, $(\pm \pi, -2a, 0)$, $(\pm \frac{\pi}{2}, 0, \pm 4a)$
Extrema	$(\pm \frac{\pi}{2}, 0, \pm 4a)$ are y-extrema
	$\left(\pm \frac{\pi}{4}, 2\sqrt{2}\,a, \pm\sqrt{2}\,a\right)$ are x-maxima
	$\left(\pm \frac{5\pi}{4}, -2\sqrt{2}\,a, \pm\sqrt{2}\,a\right)$ are x-minima
Extent	same as extrema
Symmetries	$x = 0$; $y = 0$; $(0, 0)$
Cusp	$(0, 2a, 0)$, $(\pm \pi, -2a, 0)$

Analysis of the Nephroid.

$x = 3a \cos t - a \cos 3t$

$y = 3a \sin t - a \sin 3t$

$\dot{x} = 6a \sin t \cos 2t$

$\ddot{x} = 6a \cos t\, (6 \cos^2 t - 5)$

$\dot{y} = 12a \cos t \sin^2 t$

$\ddot{y} = 12a \sin t\, (2 - 3 \sin^2 t)$

$\psi = 2t$

$s = 6a\,(1 - \cos t)$

$\rho = 3a \sin t$

6.5. Hypocycloid

The *hypocycloid* (Figure 63) is a hypotrochoid with $h = b$; the parametric equations are

6.5.1) ... $\begin{cases} x = n \cos t + b \cos \frac{n}{b} t \\ y = n \sin t - b \sin \frac{n}{b} t \end{cases}$ $-\pi \leq t \leq \pi$,

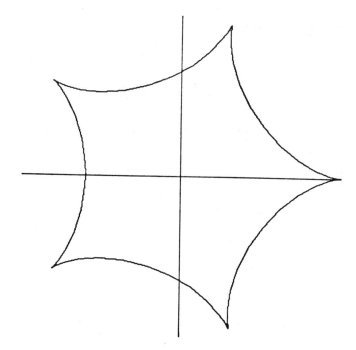

Figure 63.   Hypocycloid
b = 2; n = 8

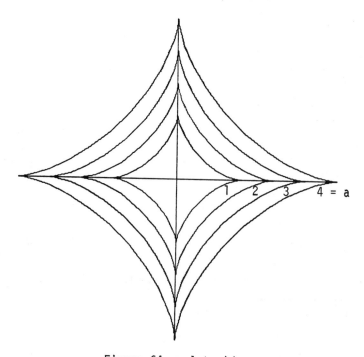

Figure 64. . Astroid
a = 1, 2, 3, 4

where $n = a - b$. If n/b is rational, the curve is algebraic and unicursal; otherwise, it is transcendental. Special cases include:

Circle	$b = 0$
Segment of a line	$a = 2b$
Deltoid	$a = 3b$
Astroid	$a = 4b$.

There are $\frac{n}{b} + 1$ cusps if n/b is an integer. The curve is symmetric about the x-axis, and is symmetric about the y-axis if n/b is an odd integer. It is completely contained within a circle defined by $|r| \leq n + b$. Its length is $L = 8n$ if n/b is an integer, and its area is $A = \pi n (2n + b)$.

6.6 Astroid (Roemer, 1674; Bernoulli, 1691)

The *astroid* (Figure 64) is a hypocycloid of four cusps, with parametric equations

6.6.1) ...
$$\begin{cases} x = a\,(3\cos t + \cos 3t) \\ y = a\,(3\sin t - \sin 3t) \end{cases} \quad -\pi \leq t \leq \pi .$$

This can also be written

6.6.2) ...
$$\begin{cases} x = 4a\cos^3 t \\ y = 4a\sin^3 t . \end{cases}$$

The Cartesian equation is

6.6.3) ... $\quad x^{2/3} + y^{2/3} = a^{2/3}$;

expanded, this becomes

6.6.4) ... $\quad (x^2 + y^2 - a^2)^3 + 27\,a^2\,x^2\,y^2 = 0$.

Other equations of interest include:

6.6.5) ... $\quad r^2 = a^2 - 3p^2 \quad$ (pedal)

6.6.6) ... $\quad s = a \cos 2\phi \quad$ (Whewell)
6.6.7) ... $\quad 4s + \rho^2 = 4a^2 \quad$ (Cesáro)
6.6.8) ... $\quad 4(\rho^2 + 4s^2) = 9a^2$.

The curve has area $A = \frac{3}{8}\pi a^2$, length $L = 6a$, surface of revolution $\Sigma_x = \frac{12}{5}\pi a^2$, and volume of revolution $V_x = \frac{32}{105}\pi a^3$.

Geometry of the Astroid.

Intercepts $\quad (0, 4a, 0), (\pm\pi, -4a, 0), (\pm\frac{\pi}{2}, 0, \pm 4a)$
Extrema $\quad$ same as intercepts
Extent $\quad$ same as intercepts
Symmetries $\quad x = 0; y = 0; (0, 0)$
Cusp $\quad$ same as intercepts

Analysis of the Astroid.

$x = 4a \cos^3 t$

$y = 4a \sin^3 t$

$\dot{x} = -12a \cos^2 t \sin t$

$\ddot{x} = -12a \cos t (1 - 3\sin^2 t)$

$\dot{y} = 12a \sin^2 t \cos t$

$\ddot{y} = 12a \sin t (3\cos^2 t - 1)$

$\theta = t$

$m = -\tan t$

$\psi = \pi - t$

$s = 6a \sin^2 t$

$\rho = 6a \sin 2t$

6.7. Rhodonea (Grandi, 1723)

The *rhodonea* (or *rose*) is the pedal of an epicycloid with respect to the center (Figures 65-66). The polar equation is

6.7.1) ... $\quad r = a \cos m\theta$.

If m is an integer, then there are m petals if m is odd, and 2 m petals if m is even. Special cases are:

$$\begin{array}{ll} \text{Trifolium} & m = 3 \\ \text{Quadrifolium} & m = 2 \\ \text{Pedal of a Cardioid} & m = 1/3 \end{array}$$

The curve is completely contained within a circle defined by $|r| < a$.

Geometry of the Rhodonea.

Intercepts	$(0, 0)$, $(a, 0)$
	$(-a, 0)$, $(0, \pm a)$ if m is even
Symmetry	$y = 0$
	$x = 0$; $(0, 0)$ if m is even
Node	$(0, 0)$

6.8. Nephroid of Freeth

Freeth's nephroid (Figure 67) is the strophoid of a circle with respect to the center as pole, fixed point on the circumference. The polar equation is

6.8.1) ... $\quad r = a (1 + 2 \sin \frac{1}{2} \theta)$.

Geometry of the Nephroid of Freeth.

Intercepts	$(a, 0)$, $(-3a, 0)$, $(0, 0)$
	$\left(0, \pm a (\sqrt{2} + 1)\right)$, $\left(0, \pm a (\sqrt{2} - 1)\right)$

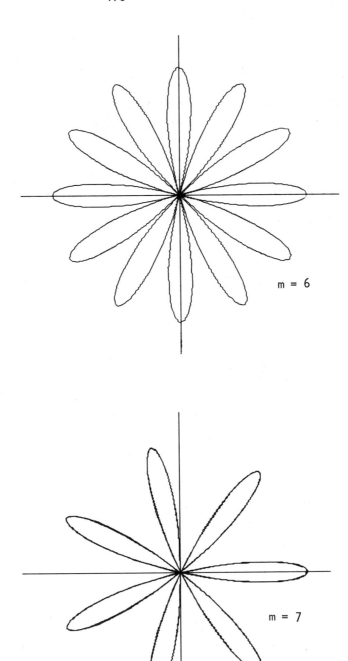

Figure 65.   Rhodonea -1
a = 4;   m = 6, 7

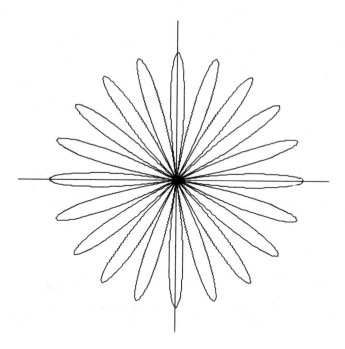

Figure 66. Rhodonea -2
a = 4; m = 10

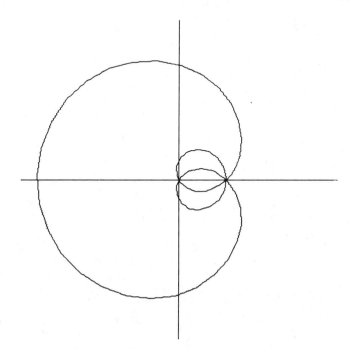

Figure 67. Nephroid of Freeth
a = 3

Extent	$\|r\| \leq 3a$
Symmetry	$x = 0$; $y = 0$; $(0, 0)$
Node	$(0, 0)$, $(a, 0)$

6.9. Cayley's Sextic (Maclaurin, 1718)

Cayley's sextic (Figure 68) is defined by the equation

$$6.9.1) \quad r = a \cos^3 \tfrac{1}{3}\theta \ .$$

It is the roulette of a cardioid with respect to an equal cardioid and the cusp. The Cartesian equation is given by

$$6.9.2) \quad 4(x^2 + y^2 - ax)^3 = 27 a^2 (x^2 + y^2)^2 \ .$$

Geometry of Cayley's Sextic.

Intercepts	$(0, 0)$, $(a, 0)$, $(-\tfrac{1}{8}a, 0)$, $\left(0, \pm \tfrac{3}{8}\sqrt{3}\, a\right)$
Extent	$\|r\| \leq a$
Symmetry	$y = 0$
Node	$(-\tfrac{1}{8}a, 0)$

6.10. Bowditch Curve (Bowditch, 1815)

The *Bowditch curve* (or the *curve of Lissajous*) is defined by the equations

$$6.10.1) \quad \begin{cases} x = a \sin(nt + d) \\ y = b \sin t \ . \end{cases}$$

It is illustrated in Figures 69-71 for eighteen selected values of n and d. The curve is algebraic and unicursal if n is rational, and transcendental otherwise. It is entirely contained within a rectangle defined by $|x| \leq a$, $|y| \leq b$.

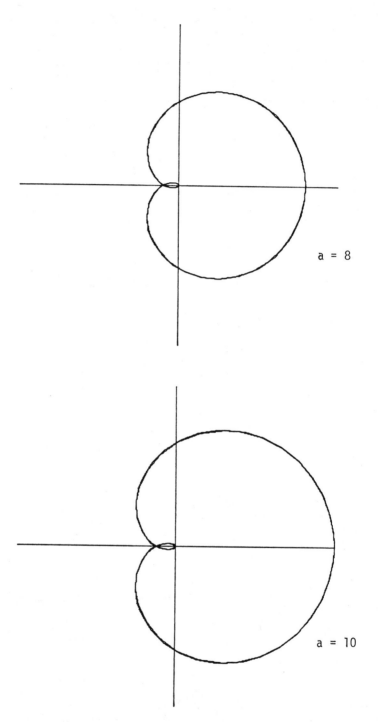

Figure 68. Cayley's Sextic.
a = 8, 10

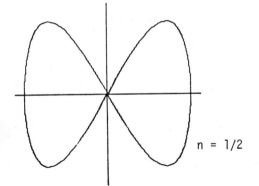
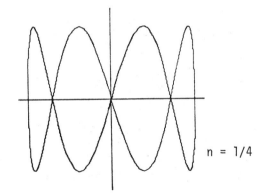
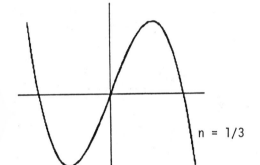
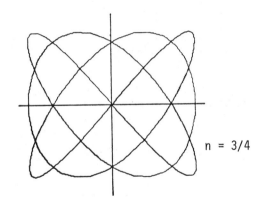
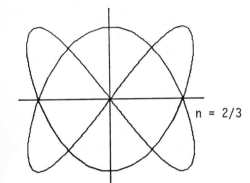
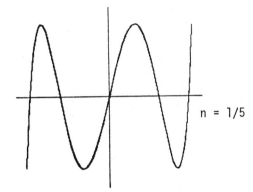

Figure 69. Bowditch Curve -1

a = 9; b = 8; d = 0; n = 1/2; 1/3, 2/3; 1/4, 3/4; 1/5

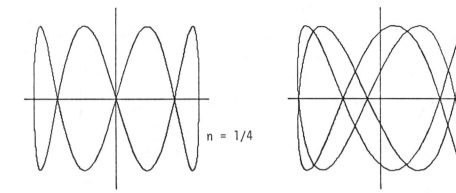
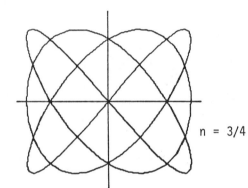
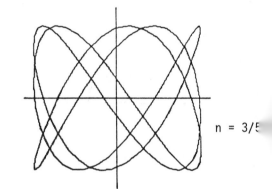
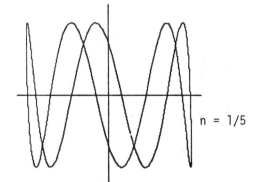
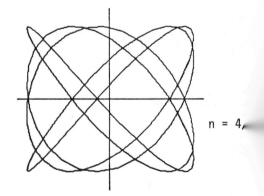

Figure 70. Bowditch Curve -2
a = 9; b = 8; d = π/4; n = 1/4, 3/4; 1/5, 2/5, 3/5, 4/5

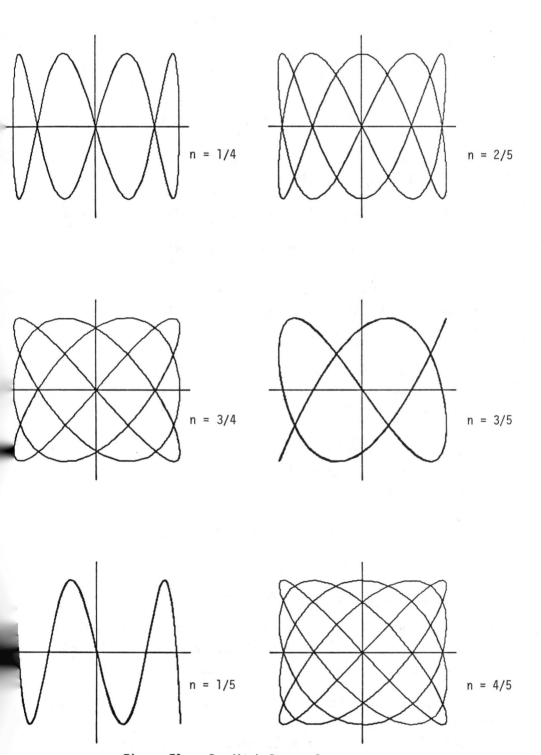

Figure 71. Bowditch Curve -3
a = 9; b = 8; d = π; n = 1/4, 3/4; 1/5, 2/5, 3/5, 4/5

CHAPTER 7

TRANSCENDENTAL CURVES

7.1. Sinusoidal Spiral (Maclaurin, 1718)

The *sinusoidal spiral* is defined as the locus of the equation

7.1.1) ... $r^n = a^n \cos n\theta$,   n rational.

Many of these curves reduce to other curves, so this spiral is not discussed here in detail. Special cases include:

	n	section
Logarithmic spiral	0	7.2
Cayley's sextet	1/3	6.9
Cardioid	1/2	5.2
Circle	1	3.2
Lemniscate of Bernoulli	2	5.3
Tschirnhausen's Cubic	- 1/3	4.2
Parabola	- 1/2	3.3
Line	- 1	---
Equilateral hyperbola	- 2	3.5

7.2. Logarithmic Spiral (Descartes, 1638)

The *logarithmic spiral* (Figure 72), also known as the *equiangular spiral* and the *logistique*, is defined as a spiral that cuts radius vectors at a constant angle ϕ. The polar equation is

7.2.1) ... $r = \exp(a\theta)$

where $a = \cot \phi$. The Cartesian equation is clearly

7.2.2) ... $x^2 + y^2 = \exp[2a \operatorname{atan}(y/x)]$.

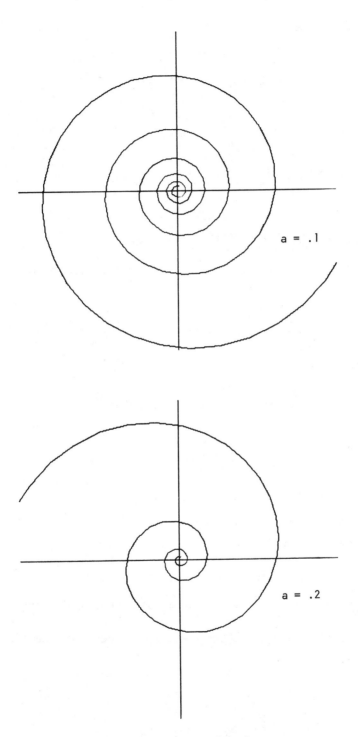

Figure 72. Logarithmic Spiral
a = .1, .2

Other equations are:

7.2.3) ... $s = r \sec \phi$ (Whewell)

7.2.4) ... $\rho = s$ (Cesáro)

7.2.5) ... $p = r \sin \phi$ (pedal).

The curve is asymptotic to the origin.

7.3. Archimedean Spirals (Sacchi, 1854)

The *Archimedean spirals* (Figures 73-77) are defined by the equation

7.3.1) ... $r^m = a^m \theta$.

The pedal equation is

7.3.2) ... $p^2 (m^2 r^{2m} + a^{2m}) = m^2 r^{2m+2}$.

Special cases are:

	m	Figure
Archimedes' spiral (Archimedes, 225 B.C.)	1	73
Fermat's spiral (Fermat, 1636)	2	74
Hyperbolic spiral (Varignon, 1704)	-1	75
Lituus (Cotés, 1722)	-2	76

The hyperbolic spiral is also known as the reciprocal spiral.

This spiral inverts into $r^{-m} = a^{-m} \theta$; thus, for example, Fermat's spiral and the lituus are inverses.

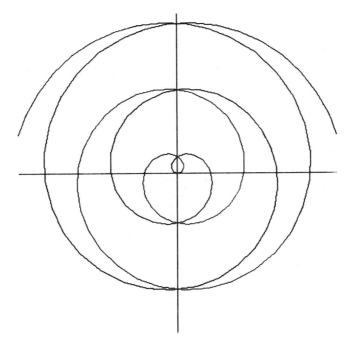

Figure 73: Archimedes' Spiral
a = 2; m = 1

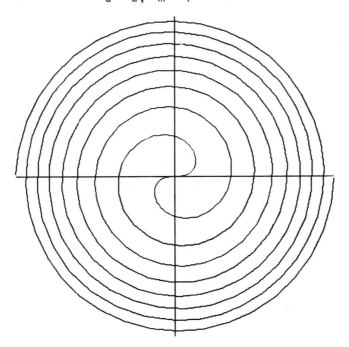

Figure 74. Fermat's Spiral
a = 6; m = 2

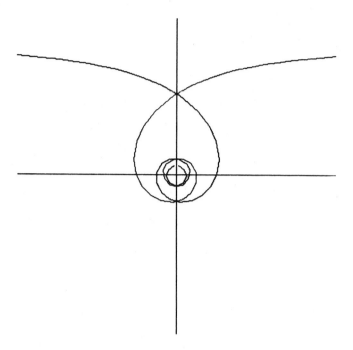

Figure 75.   Hyperbolic Spiral
$a = 2; \quad m = -1$

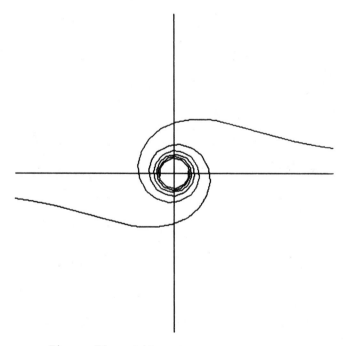

Figure 76.   Lituus
$a = 2; \quad m = -2$

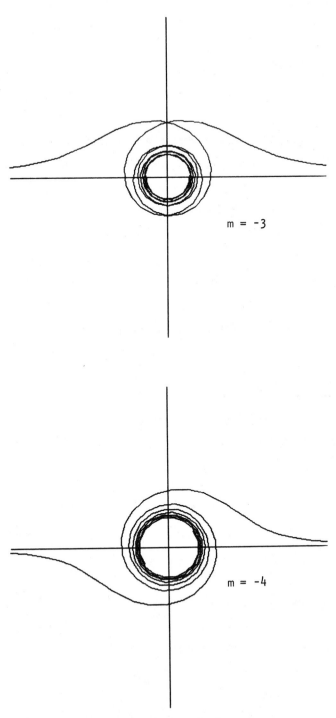

Figure 77. Archimedean Spiral
a = 2; m = -3, -4

The hyperbolic spiral has asymptote $y = a$, and the lituus has asymptote $y = 0$.

7.4. Euler's Spiral (Euler, 1744)

Euler's spiral (also known as the *clothoid*, or *spiral of Cornu*) is defined by the parametric equations

7.4.1) ...
$$\begin{cases} x = \pm a \int_0^t \frac{\sin t}{\sqrt{t}}\, dt \\ y = \pm a \int_0^t \frac{\cos t}{\sqrt{t}}\, dt \end{cases} \quad 0 \leq t < \infty.$$

These, of course, are the Fresnel integrals. The intrinsic equation is

7.4.2) ... $\quad 2\rho s = a^2$.

The curve has asymptotic points given by $(\pm a/2, \pm a/2)$. It is illustrated in Figure 78.

7.5. Involute of a Circle (Huygens, 1693)

The *involute of a circle* has equations

7.5.1) ...
$$\begin{cases} x = a(\cos t + t \sin t) \\ y = a(\sin t - t \cos t) \end{cases} \quad -\infty < t < \infty$$

(by using 2.1.5). Intrinsic equations are

7.5.2) ... $\quad 2s = a\phi^2 \quad$ (Whewell)

7.5.3) ... $\quad \rho^2 = 2as \quad$ (Cesáro).

The spiral is shown in Figure 79.

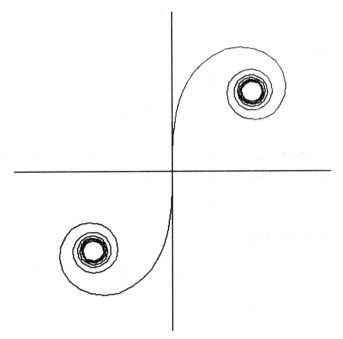

Figure 78... Euler's Spiral
a = 1

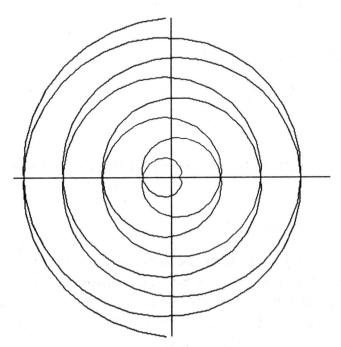

Figure 79. Involute of a Circle
a = 1

7.6. Epi Spiral

The *epi spiral* is defined by the polar equation

7.6.1) ... $\quad r \cos n\theta = a$.

There are n sections if n is an odd integer, and 2 n sections if n is an even integer (Figure 80). The curve inverts into a rose.

7.7. Poinsot's Spirals

The two *spirals of Poinsot* are defined by the polar equations

7.7.1) ... $\quad r \cosh n\theta = a$
and
7.7.2) ... $\quad r \sinh n\theta = a$.

They are shown in Figures 81 and 82, respectively.

7.8. Cochleoid (Bernoulli, 1726)

The *cochleoid* (Figure 83) is defined by the equation

7.8.1) ... $\quad r\theta = a \sin \theta$.

It has an asymptote at the origin, and intersects the x-axis at (a, 0).

7.9. Cycloid (Mersenne, 1599)

The *cycloid* is defined as the locus of a point P attached to a circle C rolling on a line. If a is the diameter of the circle, h is the distance of P from the center of the circle, and the line is y = 0, then the equation of the cycloid is

7.9.1) ... $\quad \begin{cases} x = at - h \sin t \\ y = a - h \cos t \end{cases} \quad -\infty < t < \infty$.

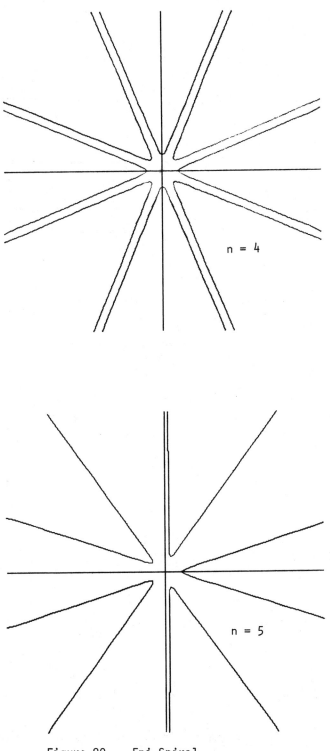

Figure 80. Epi Spiral
a = 1; n = 4, 5

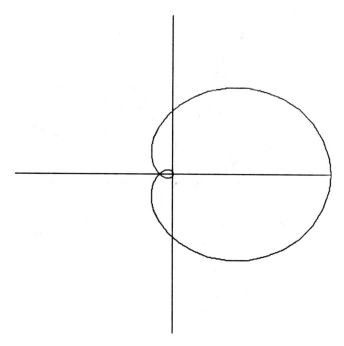

Figure 81. Poinsot's Spiral #1
a = 1; n = 1

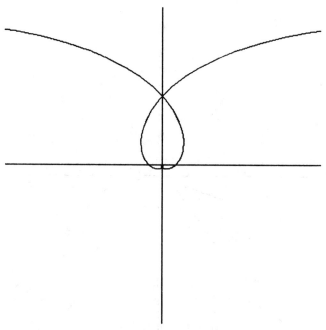

Figure 82. Poinsot's Spiral #2
a = 1; n = 1

Another equation results if the line is $y = 2a$ and the circle has center $(x, 2a)$. In this case (Figures 84-86), the equation is

7.9.2) ... $\begin{cases} x = at + h \sin t \\ y = a - h \cos t \end{cases} \quad -\infty < t < \infty .$

There are three cases. If P is on the circumference of the circle ($a = h$), the cycloid is ordinary. If P is inside the circle ($h < a$), the curve is a *curtate cycloid.* If P is outside the circle ($h > a$), the curve is a *prolate cycloid.*

Intrinsic equations are

7.9.3) ... $\rho^2 + s^2 = 16 a^2$ (Cesáro)

7.9.4) ... $s = 4a(1 - \sin \phi)$ (Whewell) .

If $a = h$, each arch has area $A = 3\pi a^2$, base line $L_x = 2\pi a$, and length $L = 8a$.

7.10. Quadratrix of Hippias (Hippias of Elis, 430 B.C.)

The *quadratrix of Hippias* is defined by the equation

7.10.1) ... $y = x \cot\left(\dfrac{\pi x}{2a}\right) .$

It is illustrated in Figure 87; the upper drawing is an expanded portion of the center of the lower drawing.

7.11. Catenary (Huygens, 1691)

The *catenary*, also known as the *chainette* and the *alysoid*, describes the form assumed by a perfect flexible inextensible chain of uniform density hanging from two supports. The equation is

7.11.1) ... $y = a \cosh\left(\dfrac{x}{a}\right) .$

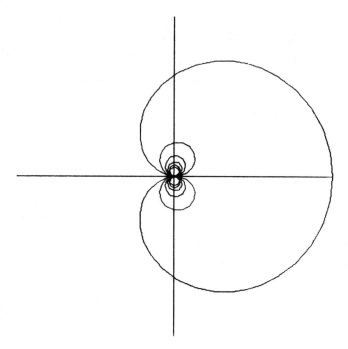

Figure 83. Cochleoid
a = 1

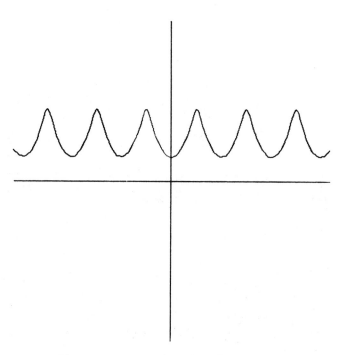

Figure 84. Curtate Cycloid
a = 1; b = 6; h = 1/2

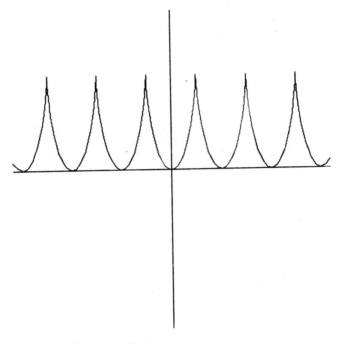

Figure 85. Cycloid
a = 1; b = 6; h = 1

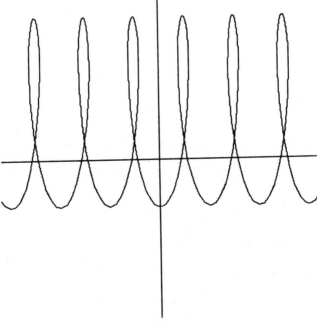

Figure 86. Prolate Cycloid
a = 1; b = 6; h = 2

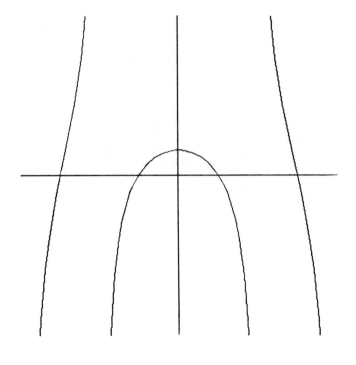

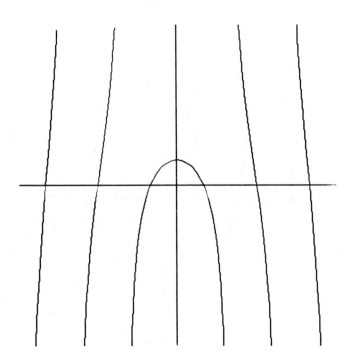

Figure 87. Quadratrix of Hippias
a = 1

Intrinsic equations are

7.11.2) ... $\quad a\rho = s^2 + a^2 \quad$ (Cesáro)

7.11.3) ... $\quad s = a \tan \phi \quad$ (Whewell)

7.11.4) ... $\quad \rho = a \sec^2 \phi$.

An illustration is given in Figure 88.

7.12. Tractrix (Huygens, 1692)

The *tractrix*, or *equitangential curve*, is the evolute of a catenary. It has equations

7.12.1) ... $\quad \begin{cases} x = a \ln (\sec t + \tan t) - a \sin t \\ y = a \cos t \end{cases} \quad -\frac{\pi}{2} < t < \frac{\pi}{2}$

7.12.2) ... $\quad x = \pm a \cosh^{-1}\left(\frac{a}{y}\right) - \sqrt{a^2 - y^2}$

7.12.3) ... $\quad \rho = a \tan \phi$

7.12.4) ... $\quad a^2 + \rho^2 = a^2 \exp(2s/a) \quad$ (Cesáro)

7.12.5) ... $\quad s = -a \ln (\sin \phi) \quad$ (Whewell) .

The curve is asymptotic to the x-axis.

The tractrix (Figure 89) represents the path of a particle P pulled by an inextensible string whose end moves along the x-axis. It has surface and volume of revolution

$$\Sigma_x = 4\pi a^2 \quad \text{and} \quad V_x = \frac{2}{3}\pi a^3 .$$

The area between the curve and the x-axis is $A = \frac{1}{2}\pi a^2$. There is a cusp at (0, a).

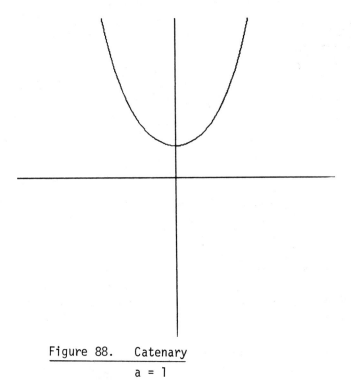

Figure 88.   Catenary
a = 1

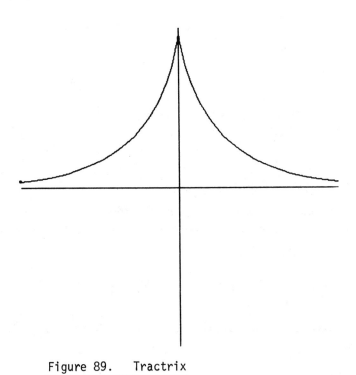

Figure 89.   Tractrix
a = 1

REFERENCES

1. E.H. Lockwood, *A Book of Curves*, Cambridge, 1961.

2. Robert C. Yates, *A Handbook on Curves and Their Properties*, J.W. Edwards, 1952.

3. Erwin Kreyszig, *Differential Geometry*, Univ. of Toronto Press, 1959.

4. L.P. Eisenhart, *Coordinate Geometry*, Dover, 1960.

5. George Salmon, *A Treatise on the Higher Plane Curves*, third edition, Chelsea. (First edition was in 1879.)

6. A.B. Basset, *An Elementary Treatise on Cubic and Quartic Curves*, Cambridge, 1901.

7. Harold Hilton, *Plane Algebraic Curves*, Oxford, 1932.

APPENDIX A

TABLES OF DERIVED CURVES

Tables 6 through 15 demonstrate the interconnections between the various special curves described in chapters 3-7.

TABLE 6. Evolutes and Involutes

Involute	Evolute
Parabola	Semi-cubical parabola
Limacon of Pascal	Catacaustic of a circle for a point source
Cardioid	Cardioid with parameter 1/3 a
Deltoid	Deltoid with parameter 3 a
Epicycloid	Epicycloid
Nephroid	Nephroid with parameter 1/2 a
Hypocycloid	Hypocycloid
Astroid	Astroid with parameter 2 a
Cayley's sextic	Nephroid
Logarithmic spiral	An equal spiral
Cycloid	An equal cycloid
Tractrix	Catenary

TABLE 7. Radials

Base Curve	Radial
Deltoid	Trifolium
Epicycloid	Rhodonea
Astroid	Quadrifolium
Logarithmic spiral	Logarithmic spiral
Cycloid	Circle
Catenary	Kampyle of Eudoxus
Tractrix	Kappa Curve

TABLE 8. Inverses

Curve	Center of inversion	Center of inversion	Inverse
Line	Not on line	On circum.	Circle
Circle	Not on circum.	Not on circum.	Circle
Parabola	Focus	Cusp	Cardioid
	Vertex	Cusp	Cissoid of Diocles
Central conic	Focus	Pole or node	Limacon
	Center	Center	Oval, Figure eight
Hyperbola (a = b)	Center	Center	Lemniscate
	Vertex	Node	Right strophoid
(a = $\sqrt{3}$ b)	Vertex	Node	Trisectrix of Maclaurin
Right strophoid	Pole	Pole	The same strophoid
Trisectrix of Maclaurin	Focus	---	Tschirnhausen's cubic
Sinusoidal spiral	Pole	Pole	Sinusoidal spiral
Logarithmic spiral	Pole	Pole	Logarithmic spiral
Archimedean spiral	Pole	Pole	Archimedean spiral
Lituus	Pole	Pole	Fermat's spiral
Epi spiral	Pole	Pole	Rhodonea
Cochleoid	Pole	---	Quadratrix of Hippias

TABLE 9. Pedals

Curve	Pole	Pedal
Line	Any point	Point
Circle	Any point	Limacon of Pascal
	On circum.	Cardioid
Parabola	On directrix	Strophoid
	Foot of directrix	Right strophoid
	Reflection of focus in directrix	Trisectrix of Maclaurin
	Vertex	Cissoid of Diocles
	Focus	Line
Central conic	Focus	Circle
	Center	Lemniscate of Bernoulli
Tschirnhausen's cubic	Focus (of pedal)	Parabola
Cissoid of Diocles	Focus	Cardioid
Cardioid	Cusp	Cayley's sextic
Deltoid	Cusp	Simple folium
	Vertex	Double folium
	Center	Trifolium
	On deltoid	Unsymm. double folium
Epicycloid	Center	Rhodonea
Hypocycloid	Center	Rhodonea
Astroid	Center	Quadrifolium
Sinusoidal spiral	Center	Sinusoidal spiral
Logarithmic spiral	Pole	An equal spiral
Involute of a circle	Center	Archimedes' spiral

TABLE 10. Strophoids

Curve	Pole	Fixed Point	Strophoid
Line	Not on line	On line	Oblique strophoid
	Not on line	Foot of Perpendicular Pole to line	Right strophoid
Circle	Center	On circum.	Nephroid of Freeth

TABLE 11. Cissoids

Curve -1	Curve -2	Pole	Cissoid
Line	Parallel line	Any point	Line
	Circle	Center	Conchoid of Nicomedes
Circle	Tangent line	On circum.	Oblique cissoid
	Tangent line	On circum., opposite tangent	Cissoid of Diocles
	Radial line	On circum.	Strophoid
	Concentric Circle	Center	Circle
	Same circle	$(a\sqrt{2}, 0)$	Lemniscate

TABLE 12. Roulettes

Curve -1	Curve -2	Pole	Roulette
Line	Any curve	On line	Involute of curve -2
Circle	Line	On circum.	Cycloid
	Circle outside C-1	On circum.	Epicycloid
	Circle inside C-1	On circum.	Hypocycloid
	Circle	Any point	Rhodonea
Parabola	Equal parabola	Vertex	Cissoid of Diocles
	Line	Focus	Catenary
Ellipse	Line	Focus	Elliptic catenary
Hyperbola	Line	Focus	Hyperbolic catenary
Logarithmic spiral	Line	Any point	Line
Hyperbolic spiral	Line	Pole	Tractrix
Involute of circle	Line	Center	Parabola
Cycloid	Line	Center	Ellipse

TABLE 13. Isoptics

Curve	Isoptic
Parabola	Hyperbola
Epicycloid	Epitrochoid
Hypocycloid	Hypotrochoid
Sinusoidal spiral	Sinusoidal spiral
Cycloid	Curtate or prolate cycloid

TABLE 14. Orthoptics

Curve	Orthoptic
Parabola	Directrix
Cardioid	Circle, limacon of Pascal
Deltoid	Circle
Astroid	Quadrifolium
Logarithmic spiral	Equal spiral

TABLE 15. Catacaustics

Curve	Source	Catacaustic
Circle	On circum	Cardioid
	Not on circum.	Limacon of Pascal
	∞	Nephroid
Parabola	Rays perp. axis	Tschirnhausen's cubic
Tschirnhausen's cubic	Focus	Semi-cubical parabola
Cissoid of Diocles	Focus	Cardioid
Cardioid	Cusp	Nephroid
Quadrifolium	Center	Astroid
Deltoid	∞	Astroid
Logarithmic Spiral	Pole	An equal spiral
Cycloid arch	Rays perp. axis	Two cycloid arches
$y = \ln x$	Rays parallel axis	Catenary

APPENDIX B

FURTHER READING

The study of special plane curves appears to grow on one. The list presented here will enable the interested reader to pursue this study. It is arranged chronologically rather than alphabetically because I have found a chronological bibliography to be more useful.

1. Cayley, Arthur, "On the classification of cubic curves," *Trans. Camb. Phil. Soc.* XI (1866) 81-128.

2. Besant, W.H., *Roulettes and Glissettes*, London (1870).

3. Salmon, George, *Lessons in Higher Algebra*, Dublin (1876).

4. Proctor, R.A., *The Geometry of Cycloids* (1878).

5. Salmon, George, *Higher Plane Curves*, Dublin (1879); reprinted by Chelsea Pub. Co. 1960.

6. Salmon, George, and O.W. Fiedler, *Analytische Geometrie der höheren ebenen Kurven*, Leipzig (1882).

7. Casey, J., *A Treatise on the Analytic Geometry of the Point, Line, Circle, and Conic Section*, London (1883).

8. Schroeter, Heinrich, *Theorie der ebenen Kurven*, Leipzig (1888).

9. Brocard, H., "Le trifolium," *Mathesis* (2) vol. 2 (1892), suppl. II.

10. Frost, P., *Curve Tracing*, Macmillan (1892); reprinted by Chelsea Pub. Co.

11. Basset, A.B., *Elementary Treatise on Cubic and Quadric Curves* (1901

12. Smith, A.C., "Certain hyperbolic curves of the nth order," *Amer. Math. Mon.* 8 (1901) 241.

13. Converse, H.A., "On a system of hypocycloids," *Annals of Math.* (2) 5 (1903-04) 105-139.

14. Grace, J.H., and A. Young, *Algebra of Invariants*, Cambridge (1903).

15. Carmichael, R.D., "On a certain class of curves given by transcendental equations," *Amer. Math. Mon.* 13 (1906), 221-226.

16. Carmichael, R.D., "On a certain quartic curve which may degenerate into an ellipse," *Amer. Math. Mon.* 14 (1907) 52.

17. Carmichael, R.D., "On a certain class of quartic curves," *Amer. Math. Mon.* 15 (1908) 7.

18. Texeira, F. Gomes, *Traité des courbes spéciales remarqueables planes et gauches* (1908-15).

19. Wieleitner, H., *Spezielle ebene Kurven*, Leipzig (1908).

20. Lefschetz, S., "On the existence of loci with given singularities," *Trans. Amer. Math. Soc.* 14 (1913) 23-41.

21. "Euler integrals and Euler's spiral — sometimes called Fresnel integrals and the clothoide or Cornu's spiral," *Amer. Math. Mon.* 25 (1918) 276-282.

22. "The logarithmic spiral," *Amer. Math. Mon.* 25 (1918) 189-193.

23. Hodge, F.H., "Relating to generalizations of the witch and the cissoid," *Amer. Math. Mon.* 25 (1918) 223-225.

24. Light, G.H.. "The existence of cusps on the evolute at points of maximum and minimum curvature on the base curve," *Amer. Math. Mon.* 26 (1919) 151-154.

25. Rigge, W.F., "Cuspidal rosettes," *Amer. Math. Mon.* 26 (1919) 332-340.

26. Merrill, A.S., "The danger area curve," *Amer. Math. Mon.* 27 (1920) 398-401.

27. Rigge, W.F., "Envelope rosettes," *Amer. Math. Mon.* 27 (1920) 151-157.

28. Morley, F.V., "A curve of pursuit," *Amer. Math. Mon.* 28 (1921) 54-61.

29. Rigge, W.F., "Cuspidal envelope rosettes," *Amer. Math. Mon.* 29 (1922) 6-8.

30. Morley, F.V., "The three bar curve," *Amer. Math. Mon.* 31 (1924) 71-7.

31. Woods, R., "The cochlioid," *Amer. Math. Mon.* 31 (1924) 222-227.

32. Ganguli, Surendramohan, *The Theory of Plane Curves*, Univ. of Calcutta (1925).

33. Brown, B.H., "La courbe du diable," *Amer. Math. Mon.* 33 (1926) 273-274.

34. Green, H.G., "The asymptotes of plane curves," *Math. Gazette* 13 (1926) 232-235.

35. Nunn, T.P., "Asymptotes," *Math. Gazette* 13 (1926) 97-103.

36. Campbell, A.O., "A simple way to discuss points of inflection on plane cubic curves," *Amer. Math. Mon.* 34 (1927) 426-429.

37. Kempner, A.J., "The devil's curve again," *Amer. Math. Mon.* 34 (1927) 262-263.

38. Burington, R.S., and H.K. Holt, "Canonical forms of plane cubic curves," *Ann. Math.* 30 (1928) 52-60.

39. Hollcroft, T.R., "Multiple points of algebraic curves," *AMS Bulletin* 35 (1929) 841-849.

40. Copeland, L.P., "Matrix conditions for multiple points of a ternary cubic," *Ann. Math.* 31 (1930) 629-632.

41. Fox, C., "The polar equations of a curve," *Math. Gazette* 15 (1930) 486-487.

42. Mallison, H.V., "The involute of the astroid," *Math. Gazette* 15 (1930) 22.

43. Neville, E.H., "The curve of pursuit," *Math. Gazette* 15 (1930) 436.

44. Coolidge, J.L., *Treatise on Algebraic Plane Curves*, Oxford (1931); reprinted by Dover (1959).

45. Bilinsky, S., "Cycloidal curves," *Amer. Math. Mon.* 39 (1932) 409-412.

46. Hilton, Harold, *Plane Algebraic Curves*, Oxford (1932).

47. Johnston, L.S., "An unusual spiral," *Amer. Math. Mon.* 40 (1933) 596-598.

48. Neville, E.H., "The tracing of cubic curves," *Math. Gazette* 18 (1934) 258-266.

49. Arang, D., "Note on the three-cusped hypocycloid," *Math. Gazette* 21 (1937) 418.

50. Johnston, L.S., "The quadratrix and the associated cochleoid," *Amer. Math. Mon.* 44 (1937) 167-168.

51. Butchart, J.H., "The deltoid regarded as the envelope of Simson lines," *Amer. Math. Mon.* 46 (1939) 85-86.

52. Goormaghtigh, R., "On the three-cusped hypocycloid," *Math. Gazette* 23 (1939) 92.

53. Patterson, B.C., "The triangle; its deltoids and foliates," *Amer. Math. Mon.* 47 (1940) 11-18.

54. Caldwell, J.H., "Repeating curves," *Math. Gazette* 25 (1941) 165-167.

55. McCarthy, J.P., "The cissoid of Diocles," *Math. Gazette* 25 (1941) 12-15.

56. Chaundy, T.W., "The inflexions of a plane cubic," *Math. Gazette*, 26 (1942) 214.

57. Weaver, J.H., "On the cubic of Tschirnhausen," *Nat. Math. Mag.* 16 (1942) 371-374.

58. Green, H.G., "On some general ovals of Cassinian type," *Math. Gazette* 27 (1943) 4-12.

59. Green, H.G., "Real quadrics and the circle at infinity," *Math. Gazette* 27 (1943) 194-197.

60. Butchart, J.H., "Some properties of the limacon and cardioid," *Amer. Math. Mon.* 52 (1945) 384-387.

61. Hadamard, J., "On the three-cusped hypocycloid," *Math. Gazette* 29 (1945) 66-67.

62. McCarthy, J.P., "The limacon and the cardioid," *Math. Gazette* 29 (1945) 219-220.

63. Goormaghtigh, R., "On the three-cusped hypocycloid," *Math. Gazette* 30 (1946) 145-146.

64. Bilo, J., "Remarkable cubic curves in metrically special homaloidal nets," [Dutch] *Simon Stevin* 25 (1947) 59-82.

65. Bosteels, G., "The scyphoid, a rational quartic curve," [Dutch] *Nieuw. Tijdschr. Wiskunda* 35 (1947) 121-125.

66. Lilley, S., "On the construction of algebraic curve branches of given composition," *J. London Math. Soc.* 22 (1947) 67-74.

67. Noble, C.E., "An anallagmatic cubic," *Amer. Math. Mon.* 55 (1948) 7-14.

68. Aude, H.T.R., "Note on quartic curves," *Amer. Math. Mon.* 56 (1949) 165.

69. Godeaux, Lucien, *Géométrie Algébrique*, Sciences et Lettres, Liége (1949).

70. Semple, J.G., and L. Roth, *Introduction to Algebraic Geometry*, Oxford (1949).

71. Edge, W.L., "A plane quartic with eight undulations," *Proc. Edinburgh Math. Soc.* 2 (1950) 147-162.

72. Walker, R.J., *Algebraic Curves*, Dover (1950).

73. Lyness, R.C., "The Cartesian oval," *Math. Gazette* 36 (1952) 51-53.

74. Simpson, H., "On a special type of plane trinodal quartic curve," *Math. Gazette* 36 (1952) 207-208.

75. Yates, R.C., *Handbook on Curves and Their Properties*, Ann Arbor (1952).

76. Puckette, C.C., "The curve of pursuit," *Math. Gazette* 37 (1953) 256-260.

77. Edge, W.L., "A plane quartic curve with twelve undulations," *Edinburgh Math. Notes* 35 (1954) 10-13.

78. Granát, Luděk, and Miroslav Fiedler, "Rational curves with a maximum number of nodes," [Czech] *Časopis Pěst. Mat.* 79 (1954) 157-161.

79. Primrose, E.J.F., "Coincidence points of a curve, *Tôhoku Math. J.* 2 (1954) 35-37.

80. Primrose, E.J.F., "A property of quartic curves with two cusps and one node," *Edinburgh Math. Notes* 39 (1954) 1-3.

81. Northcott, D.G., "A note on the genus formula for plane curves," *J. London Math. Soc.* 30 (1955) 376-382.

82. Primrose, E.J.F., *Plane Algebraic Curves*, New York (1955).

83. McMillan, R.H., "Curves of pursuit," *Math. Gazette* 40 (1956) 1-4.

84. Lasley, J.W., "On degenerate conics," *Amer. Math. Mon.* 64 (1957) 362-364.

85. Lockwood, E.H., "Negative pedals of the ellipse," *Math. Gazette* 41 (1957) 254-257.

86. McCarthy, J.P., "The cissoid of Diocles," *Math. Gazette* 41 (1957) 102-105.

87. Bernhart, Arthur, "Curves of pursuit - II," *Scripta Math.* 23 (1958) 49-65.

88. Blum, Richard, "On a pointwise construction of the lemniscate," *Canadian Math. Bull.* 1 (1958) 1-4.

89. Good, I.J., "Pursuit curves and mathematical art," *Math. Gazette* 4 (1959) 34-35.

90. Baker, Henry F., *Principles of Geometry. Vol. 5: Analytic Principles of the Theory of Curves*, New York (1960).

91. Spain, Barry, *Analytical Quadrics*, New York (1960).

92. Lockwood, E.H., *A Book of Curves*, Cambridge, England (1961).

93. Zwikker, C., *The Advanced Geometry of Plane Curves and Their Applications*, Dover (1963).

94. Jeffrey, A., "On the determination of the envelope of a family of epitrochoids with applications," *SIAM J. Applied Math.* 14 (1966) 502-510.

95. "Curves" and "Special Curves," *Encyclopedia Britannica*.

INDEX OF CURVE NAMES

Agnesi, witch of, 90-92
Alysoid, 195
Anallagmatic, 46
Archimedean spiral, 186-190
Astroid, 4-5, 34-35, 173-174
Bernoulli, lemniscate of, 4-5, 121-123, 145, 153, 184
Bicorn, 147-149
Bowditch curve, 178-179
Bullet nose, 128-130
Cardioid, 5, 12, 113, 118-121, 168, 175, 178, 184
Cartesian oval, 5, 155-157
Cassinian ellipse, 153
Cassinian oval, 5, 153-155
Catacaustic, 60
Catalan, trisectrix of, 88
Catenary, 5, 195-199
Caustic, 60
Cayley's sextic, 178, 184
Chainette, 195
Circle, 4-5, 31, 65-66, 98, 113, 131-132, 149, 160-161, 165, 173, 175, 184, 190, 192
Circle, involute of, 190
Cissoid, 53-56, 98
Cissoid, oblique, 98
Cissoid of Diocles, 95, 98-100
Clothoid, 190
Cochleoid, 192
Cochloid, 137
Conchoid, 36, 49-51, 113, 137
Conchoid, Dürer's, 157-159
Conchoid of Nicomedes, 137-139

Cornu's spiral, 190
Cross curve, 130-131
Curtate cycloid, 195
Cycloid, 5, 192-195
Cycloid, curtate, 195
Cycloid, prolate, 195
Deltoid, 131-134, 151, 173
Descartes, folium of, 106-108
Descartes, oval of, 5, 155-157
Descartes, parabola of, 110
Devil on two sticks, 151
Devil's curve, 151
Diacaustic, 60
Diocles, cissoid of, 95, 98-100
Dürer's conchoid, 157-159
Eight curve, 124-125
Ellipse, 4-5, 14, 28-29, 34-36, 62-72-78, 130, 155, 168
Ellipse, Cassinian, 153
Epi spiral, 192
Epicycloid, 161, 168-170, 175
Epitrochoid, 113, 160-161, 168
Equiangular spiral, 184
Equitangential curve, 199
Eudoxus, kampyle of, 141-142
Euler's spiral, 190
Evolute, 40, 60, 85, 199
Fermat's spiral, 31, 186
Folium, 151-153
Folium of Descartes, 106-108
Freeth's nephroid, 175-178
Gerono, lemniscate of, 124
Glissette, 57

Gutschoven's curve, 139
Hippias, quadratrix of, 195
Hippopede, 145-146
Horopter, 111
Horse fetter, 145
l'Hospital's cubic, 88
Hyperbola, 5, 29, 31-32, 35,
    62-65, 79-82, 121, 128,
    155, 184
Hyperbolic lemniscate, 121
Hyperbolic spiral, 31, 35, 186, 190
Hypocycloid, 168, 171-173
Hypotrochoid, 165-168
Inverse, 43-46, 121, 186
Involute, 41-42, 190
Involute of a circle, 190
Isochrone, 85
Isoptic, 58-59
Kampyle of Eudoxus, 141-142
Kappa curve, 139-141
Lemniscate, hyperbolic, 121
Lemniscate of Bernoulli, 4-5,
    121-123, 145, 153, 184
Lemniscate of Gerono, 124
Limacon of Pascal, 113-118, 155, 161
Line, 4-6, 55-56, 98, 100, 137, 149
    173, 184
Lissajous, curve of, 178
Lituus, 186
Logarithmic spiral, 184-186
Logistique, 184
Maclaurin, trisectrix of, 36, 95,
    104-106
Nephroid, 168, 170-171
Nephroid of Freeth, 175-178
Newton, trident of, 110
Nicomedes, conchoid of, 137-139

Normal pedal curve, 47
Oblique cissoid, 98
Oblique strophoid, 100
Orthoptic, 58
Orthotomic, 60
Oval of Cassini, 5, 153-155
Oval of Descartes, 5, 155-157
Parabola, 4, 29, 35, 62-64, 67-72,
    184
Parabola of Descartes, 110
Parabola, semi-cubical, 35-36, 85-86,
    89
Parallel, 36, 42-43
Pascal, limacon of, 113-118, 155, 161
Pear-shaped quartic, 149
Pedal, 35, 46-49, 94-96, 100, 104
    151, 175
Piriform, 149-150
Poinsot, spiral of, 192
Power function, 30-31, 83-84
Prolate cycloid, 195
Quadratrix of Hippias, 195
Quadrifolium, 175
Radial, 40
Reciprocal spiral, 186
Rhodonea, 168, 175
Right strophoid, 95, 100-104
Rose, 175, 192
Roulette, 56-57, 113, 131, 160, 165,
    178, 192
Semi-cubical parabola, 35-36, 85-86,
    89
Serpentine, 111
Sinusoidal spiral, 88, 121, 184
Spiral, Archimedes', 186-190
Spiral, Cornu's, 190
Spiral, epi, 192

Spiral, equiangular, 184
Spiral, Euler's, 190
Spiral, Fermat's, 31, 186
Spiral, hyperbolic, 31, 35, 186, 190
Spiral, logarithmic, 184-186
Spiral, Poinsot's, 192
Spiral, reciprocal, 186
Spiral, sinusoidal, 88, 121, 184
Strophoid, 51-53, 95, 100-104, 175
Strophoid, oblique, 100
Tractrix, 5, 199
Tricuspid, 131
Trident of Newton, 110
Trifolium, 153, 175
Trisectrix, 113
Trisectrix of Catalan, 88
Trisectrix of Maclaurin, 36, 95, 104-106
Tschirnhausen's cubic, 86-90, 184
Versiera, 90
Witch of Agnesi, 90-92

A CATALOGUE OF SELECTED DOVER BOOKS
IN ALL FIELDS OF INTEREST

A CATALOGUE OF SELECTED DOVER BOOKS IN ALL FIELDS OF INTEREST

CELESTIAL OBJECTS FOR COMMON TELESCOPES, T. W. Webb. The most used book in amateur astronomy: inestimable aid for locating and identifying nearly 4,000 celestial objects. Edited, updated by Margaret W. Mayall. 77 illustrations. Total of 645pp. 5⅜ x 8½.
20917-2, 20918-0 Pa., Two-vol. set $9.00

HISTORICAL STUDIES IN THE LANGUAGE OF CHEMISTRY, M. P. Crosland. The important part language has played in the development of chemistry from the symbolism of alchemy to the adoption of systematic nomenclature in 1892. ". . . wholeheartedly recommended,"—Science. 15 illustrations. 416pp. of text. 5⅝ x 8¼. 63702-6 Pa. $6.00

BURNHAM'S CELESTIAL HANDBOOK, Robert Burnham, Jr. Thorough, readable guide to the stars beyond our solar system. Exhaustive treatment, fully illustrated. Breakdown is alphabetical by constellation: Andromeda to Cetus in Vol. 1; Chamaeleon to Orion in Vol. 2; and Pavo to Vulpecula in Vol. 3. Hundreds of illustrations. Total of about 2000pp. 6⅛ x 9¼.
23567-X, 23568-8, 23673-0 Pa., Three-vol. set $27.85

THEORY OF WING SECTIONS: INCLUDING A SUMMARY OF AIRFOIL DATA, Ira H. Abbott and A. E. von Doenhoff. Concise compilation of subatomic aerodynamic characteristics of modern NASA wing sections, plus description of theory. 350pp. of tables. 693pp. 5⅜ x 8½.
60586-8 Pa. $8.50

DE RE METALLICA, Georgius Agricola. Translated by Herbert C. Hoover and Lou H. Hoover. The famous Hoover translation of greatest treatise on technological chemistry, engineering, geology, mining of early modern times (1556). All 289 original woodcuts. 638pp. 6¾ x 11.
60006-8 Clothbd. $17.95

THE ORIGIN OF CONTINENTS AND OCEANS, Alfred Wegener. One of the most influential, most controversial books in science, the classic statement for continental drift. Full 1966 translation of Wegener's final (1929) version. 64 illustrations. 246pp. 5⅜ x 8½. 61708-4 Pa. $4.50

THE PRINCIPLES OF PSYCHOLOGY, William James. Famous long course complete, unabridged. Stream of thought, time perception, memory, experimental methods; great work decades ahead of its time. Still valid, useful; read in many classes. 94 figures. Total of 1391pp. 5⅜ x 8½.
20381-6, 20382-4 Pa., Two-vol. set $13.00

CATALOGUE OF DOVER BOOKS

DRAWINGS OF WILLIAM BLAKE, William Blake. 92 plates from Book of Job, *Divine Comedy, Paradise Lost,* visionary heads, mythological figures, Laocoon, etc. Selection, introduction, commentary by Sir Geoffrey Keynes. 178pp. 8⅛ x 11. 22303-5 Pa. $4.00

ENGRAVINGS OF HOGARTH, William Hogarth. 101 of Hogarth's greatest works: *Rake's Progress, Harlot's Progress, Illustrations for Hudibras, Before and After, Beer Street and Gin Lane,* many more. Full commentary. 256pp. 11 x 13¾. 22479-1 Pa. $12.95

DAUMIER: 120 GREAT LITHOGRAPHS, Honore Daumier. Wide-ranging collection of lithographs by the greatest caricaturist of the 19th century. Concentrates on eternally popular series on lawyers, on married life, on liberated women, etc. Selection, introduction, and notes on plates by Charles F. Ramus. Total of 158pp. 9⅜ x 12¼. 23512-2 Pa. $6.00

DRAWINGS OF MUCHA, Alphonse Maria Mucha. Work reveals draftsman of highest caliber: studies for famous posters and paintings, renderings for book illustrations and ads, etc. 70 works, 9 in color; including 6 items not drawings. Introduction. List of illustrations. 72pp. 9⅜ x 12¼. (Available in U.S. only) 23672-2 Pa. $4.00

GIOVANNI BATTISTA PIRANESI: DRAWINGS IN THE PIERPONT MORGAN LIBRARY, Giovanni Battista Piranesi. For first time ever all of Morgan Library's collection, world's largest. 167 illustrations of rare Piranesi drawings—archeological, architectural, decorative and visionary. Essay, detailed list of drawings, chronology, captions. Edited by Felice Stampfle. 144pp. 9⅜ x 12¼. 23714-1 Pa. $7.50

NEW YORK ETCHINGS (1905-1949), John Sloan. All of important American artist's N.Y. life etchings. 67 works include some of his best art; also lively historical record—Greenwich Village, tenement scenes. Edited by Sloan's widow. Introduction and captions. 79pp. 8⅜ x 11¼. 23651-X Pa. $4.00

CHINESE PAINTING AND CALLIGRAPHY: A PICTORIAL SURVEY, Wan-go Weng. 69 fine examples from John M. Crawford's matchless private collection: landscapes, birds, flowers, human figures, etc., plus calligraphy. Every basic form included: hanging scrolls, handscrolls, album leaves, fans, etc. 109 illustrations. Introduction. Captions. 192pp. 8⅞ x 11¾. 23707-9 Pa. $7.95

DRAWINGS OF REMBRANDT, edited by Seymour Slive. Updated Lippmann, Hofstede de Groot edition, with definitive scholarly apparatus. All portraits, biblical sketches, landscapes, nudes, Oriental figures, classical studies, together with selection of work by followers. 550 illustrations. Total of 630pp. 9⅛ x 12¼. 21485-0, 21486-9 Pa., Two-vol. set $15.00

THE DISASTERS OF WAR, Francisco Goya. 83 etchings record horrors of Napoleonic wars in Spain and war in general. Reprint of 1st edition, plus 3 additional plates. Introduction by Philip Hofer. 97pp. 9⅜ x 8¼. 21872-4 Pa. $4.00

CATALOGUE OF DOVER BOOKS

THE SENSE OF BEAUTY, George Santayana. Masterfully written discussion of nature of beauty, materials of beauty, form, expression; art, literature, social sciences all involved. 168pp. 5⅜ x 8½. 20238-0 Pa. $3.00

ON THE IMPROVEMENT OF THE UNDERSTANDING, Benedict Spinoza. Also contains *Ethics, Correspondence,* all in excellent R. Elwes translation. Basic works on entry to philosophy, pantheism, exchange of ideas with great contemporaries. 402pp. 5⅜ x 8½. 20250-X Pa. $4.50

THE TRAGIC SENSE OF LIFE, Miguel de Unamuno. Acknowledged masterpiece of existential literature, one of most important books of 20th century. Introduction by Madariaga. 367pp. 5⅜ x 8½.
20257-7 Pa. $4.50

THE GUIDE FOR THE PERPLEXED, Moses Maimonides. Great classic of medieval Judaism attempts to reconcile revealed religion (Pentateuch, commentaries) with Aristotelian philosophy. Important historically, still relevant in problems. Unabridged Friedlander translation. Total of 473pp. 5⅜ x 8½. 20351-4 Pa. $6.00

THE I CHING (THE BOOK OF CHANGES), translated by James Legge. Complete translation of basic text plus appendices by Confucius, and Chinese commentary of most penetrating divination manual ever prepared. Indispensable to study of early Oriental civilizations, to modern inquiring reader. 448pp. 5⅜ x 8½. 21062-6 Pa. $5.00

THE EGYPTIAN BOOK OF THE DEAD, E. A. Wallis Budge. Complete reproduction of Ani's papyrus, finest ever found. Full hieroglyphic text, interlinear transliteration, word for word translation, smooth translation. Basic work, for Egyptology, for modern study of psychic matters. Total of 533pp. 6½ x 9¼. (Available in U.S. only) 21866-X Pa. $5.95

THE GODS OF THE EGYPTIANS, E. A. Wallis Budge. Never excelled for richness, fullness: all gods, goddesses, demons, mythical figures of Ancient Egypt; their legends, rites, incarnations, variations, powers, etc. Many hieroglyphic texts cited. Over 225 illustrations, plus 6 color plates. Total of 988pp. 6⅛ x 9¼. (Available in U.S. only)
22055-9, 22056-7 Pa., Two-vol. set $16.00

THE STANDARD BOOK OF QUILT MAKING AND COLLECTING, Marguerite Ickis. Full information, full-sized patterns for making 46 traditional quilts, also 150 other patterns. Quilted cloths, lame, satin quilts, etc. 483 illustrations. 273pp. 6⅞ x 9⅝. 20582-7 Pa. $4.95

CORAL GARDENS AND THEIR MAGIC, Bronsilaw Malinowski. Classic study of the methods of tilling the soil and of agricultural rites in the Trobriand Islands of Melanesia. Author is one of the most important figures in the field of modern social anthropology. 143 illustrations. Indexes. Total of 911pp. of text. 5⅝ x 8¼. (Available in U.S. only)
23597-1 Pa. $12.95

CATALOGUE OF DOVER BOOKS

THE PHILOSOPHY OF HISTORY, Georg W. Hegel. Great classic of Western thought develops concept that history is not chance but a rational process, the evolution of freedom. 457pp. 5⅜ x 8½. 20112-0 Pa. $4.50

LANGUAGE, TRUTH AND LOGIC, Alfred J. Ayer. Famous, clear introduction to Vienna, Cambridge schools of Logical Positivism. Role of philosophy, elimination of metaphysics, nature of analysis, etc. 160pp. 5⅜ x 8½. (Available in U.S. only) 20010-8 Pa. $2.00

A PREFACE TO LOGIC, Morris R. Cohen. Great City College teacher in renowned, easily followed exposition of formal logic, probability, values, logic and world order and similar topics; no previous background needed. 209pp. 5⅜ x 8½. 23517-3 Pa. $3.50

REASON AND NATURE, Morris R. Cohen. Brilliant analysis of reason and its multitudinous ramifications by charismatic teacher. Interdisciplinary, synthesizing work widely praised when it first appeared in 1931. Second (1953) edition. Indexes. 496pp. 5⅜ x 8½. 23633-1 Pa. $6.50

AN ESSAY CONCERNING HUMAN UNDERSTANDING, John Locke. The only complete edition of enormously important classic, with authoritative editorial material by A. C. Fraser. Total of 1176pp. 5⅜ x 8½.
20530-4, 20531-2 Pa., Two-vol. set $16.00

HANDBOOK OF MATHEMATICAL FUNCTIONS WITH FORMULAS, GRAPHS, AND MATHEMATICAL TABLES, edited by Milton Abramowitz and Irene A. Stegun. Vast compendium: 29 sets of tables, some to as high as 20 places. 1,046pp. 8 x 10½. 61272-4 Pa. $14.95

MATHEMATICS FOR THE PHYSICAL SCIENCES, Herbert S. Wilf. Highly acclaimed work offers clear presentations of vector spaces and matrices, orthogonal functions, roots of polynomial equations, conformal mapping, calculus of variations, etc. Knowledge of theory of functions of real and complex variables is assumed. Exercises and solutions. Index. 284pp. 5⅝ x 8¼. 63635-6 Pa. $5.00

THE PRINCIPLE OF RELATIVITY, Albert Einstein et al. Eleven most important original papers on special and general theories. Seven by Einstein, two by Lorentz, one each by Minkowski and Weyl. All translated, unabridged. 216pp. 5⅜ x 8½. 60081-5 Pa. $3.50

THERMODYNAMICS, Enrico Fermi. A classic of modern science. Clear, organized treatment of systems, first and second laws, entropy, thermodynamic potentials, gaseous reactions, dilute solutions, entropy constant. No math beyond calculus required. Problems. 160pp. 5⅜ x 8½. 60361-X Pa. $3.00

ELEMENTARY MECHANICS OF FLUIDS, Hunter Rouse. Classic undergraduate text widely considered to be far better than many later books. Ranges from fluid velocity and acceleration to role of compressibility in fluid motion. Numerous examples, questions, problems. 224 illustrations. 376pp. 5⅝ x 8¼. 63699-2 Pa. $5.00

CATALOGUE OF DOVER BOOKS

THE COMPLETE BOOK OF DOLL MAKING AND COLLECTING, Catherine Christopher. Instructions, patterns for dozens of dolls, from rag doll on up to elaborate, historically accurate figures. Mould faces, sew clothing, make doll houses, etc. Also collecting information. Many illustrations. 288pp. 6 x 9. 22066-4 Pa. $4.50

THE DAGUERREOTYPE IN AMERICA, Beaumont Newhall. Wonderful portraits, 1850's townscapes, landscapes; full text plus 104 photographs. The basic book. Enlarged 1976 edition. 272pp. 8¼ x 11¼. 23322-7 Pa. $7.95

CRAFTSMAN HOMES, Gustav Stickley. 296 architectural drawings, floor plans, and photographs illustrate 40 different kinds of "Mission-style" homes from *The Craftsman* (1901-16), voice of American style of simplicity and organic harmony. Thorough coverage of Craftsman idea in text and picture, now collector's item. 224pp. 8⅛ x 11. 23791-5 Pa. $6.00

PEWTER-WORKING: INSTRUCTIONS AND PROJECTS, Burl N. Osborn. & Gordon O. Wilber. Introduction to pewter-working for amateur craftsman. History and characteristics of pewter; tools, materials, step-by-step instructions. Photos, line drawings, diagrams. Total of 160pp. 7⅞ x 10¾. 23786-9 Pa. $3.50

THE GREAT CHICAGO FIRE, edited by David Lowe. 10 dramatic, eye-witness accounts of the 1871 disaster, including one of the aftermath and rebuilding, plus 70 contemporary photographs and illustrations of the ruins—courthouse, Palmer House, Great Central Depot, etc. Introduction by David Lowe. 87pp. 8¼ x 11. 23771-0 Pa. $4.00

SILHOUETTES: A PICTORIAL ARCHIVE OF VARIED ILLUSTRATIONS, edited by Carol Belanger Grafton. Over 600 silhouettes from the 18th to 20th centuries include profiles and full figures of men and women, children, birds and animals, groups and scenes, nature, ships, an alphabet. Dozens of uses for commercial artists and craftspeople. 144pp. 8⅜ x 11¼. 23781-8 Pa. $4.50

ANIMALS: 1,419 COPYRIGHT-FREE ILLUSTRATIONS OF MAMMALS, BIRDS, FISH, INSECTS, ETC., edited by Jim Harter. Clear wood engravings present, in extremely lifelike poses, over 1,000 species of animals. One of the most extensive copyright-free pictorial sourcebooks of its kind. Captions. Index. 284pp. 9 x 12. 23766-4 Pa. $8.95

INDIAN DESIGNS FROM ANCIENT ECUADOR, Frederick W. Shaffer. 282 original designs by pre-Columbian Indians of Ecuador (500-1500 A.D.). Designs include people, mammals, birds, reptiles, fish, plants, heads, geometric designs. Use as is or alter for advertising, textiles, leathercraft, etc. Introduction. 95pp. 8¾ x 11¼. 23764-8 Pa. $3.50

SZIGETI ON THE VIOLIN, Joseph Szigeti. Genial, loosely structured tour by premier violinist, featuring a pleasant mixture of reminiscences, insights into great music and musicians, innumerable tips for practicing violinists. 385 musical passages. 256pp. 5⅝ x 8¼. 23763-X Pa. $4.00

TONE POEMS, SERIES II: TILL EULENSPIEGELS LUSTIGE STREICHE, ALSO SPRACH ZARATHUSTRA, AND EIN HELDENLEBEN, Richard Strauss. Three important orchestral works, including very popular *Till Eulenspiegel's Marry Pranks,* reproduced in full score from original editions. Study score. 315pp. 9⅜ x 12¼. (Available in U.S. only) 23755-9 Pa. $8.95

TONE POEMS, SERIES I: DON JUAN, TOD UND VERKLARUNG AND DON QUIXOTE, Richard Strauss. Three of the most often performed and recorded works in entire orchestral repertoire, reproduced in full score from original editions. Study score. 286pp. 9⅜ x 12¼. (Available in U.S. only) 23754-0 Pa. $7.50

11 LATE STRING QUARTETS, Franz Joseph Haydn. The form which Haydn defined and "brought to perfection." (*Grove's*). 11 string quartets in complete score, his last and his best. The first in a projected series of the complete Haydn string quartets. Reliable modern Eulenberg edition, otherwise difficult to obtain. 320pp. 8⅜ x 11¼. (Available in U.S. only) 23753-2 Pa. $7.50

FOURTH, FIFTH AND SIXTH SYMPHONIES IN FULL SCORE, Peter Ilyitch Tchaikovsky. Complete orchestral scores of Symphony No. 4 in F Minor, Op. 36; Symphony No. 5 in E Minor, Op. 64; Symphony No. 6 in B Minor, "Pathetique," Op. 74. Bretikopf & Hartel eds. Study score. 480pp. 9⅜ x 12¼. 23861-X Pa. $10.95

THE MARRIAGE OF FIGARO: COMPLETE SCORE, Wolfgang A. Mozart. Finest comic opera ever written. Full score, not to be confused with piano renderings. Peters edition. Study score. 448pp. 9⅜ x 12¼. (Available in U.S. only) 23751-6 Pa. $11.95

"IMAGE" ON THE ART AND EVOLUTION OF THE FILM, edited by Marshall Deutelbaum. Pioneering book brings together for first time 38 groundbreaking articles on early silent films from *Image* and 263 illustrations newly shot from rare prints in the collection of the International Museum of Photography. A landmark work. Index. 256pp. 8¼ x 11. 23777-X Pa. $8.95

AROUND-THE-WORLD COOKY BOOK, Lois Lintner Sumption and Marguerite Lintner Ashbrook. 373 cooky and frosting recipes from 28 countries (America, Austria, China, Russia, Italy, etc.) include Viennese kisses, rice wafers, London strips, lady fingers, hony, sugar spice, maple cookies, etc. Clear instructions. All tested. 38 drawings. 182pp. 5⅜ x 8. 23802-4 Pa. $2.50

THE ART NOUVEAU STYLE, edited by Roberta Waddell. 579 rare photographs, not available elsewhere, of works in jewelry, metalwork, glass, ceramics, textiles, architecture and furniture by 175 artists—Mucha, Seguy, Lalique, Tiffany, Gaudin, Hohlwein, Saarinen, and many others. 288pp. 8⅜ x 11¼. 23515-7 Pa. $6.95

CATALOGUE OF DOVER BOOKS

THE AMERICAN SENATOR, Anthony Trollope. Little known, long unavailable Trollope novel on a grand scale. Here are humorous comment on American vs. English culture, and stunning portrayal of a heroine/villainess. Superb evocation of Victorian village life. 561pp. 5⅜ x 8½.
23801-6 Pa. $6.00

WAS IT MURDER? James Hilton. The author of *Lost Horizon* and *Goodbye, Mr. Chips* wrote one detective novel (under a pen-name) which was quickly forgotten and virtually lost, even at the height of Hilton's fame. This edition brings it back—a finely crafted public school puzzle resplendent with Hilton's stylish atmosphere. A thoroughly English thriller by the creator of Shangri-la. 252pp. 5⅜ x 8. (Available in U.S. only)
23774-5 Pa. $3.00

CENTRAL PARK: A PHOTOGRAPHIC GUIDE, Victor Laredo and Henry Hope Reed. 121 superb photographs show dramatic views of Central Park: Bethesda Fountain, Cleopatra's Needle, Sheep Meadow, the Blockhouse, plus people engaged in many park activities: ice skating, bike riding, etc. Captions by former Curator of Central Park, Henry Hope Reed, provide historical view, changes, etc. Also photos of N.Y. landmarks on park's periphery. 96pp. 8½ x 11.
23750-8 Pa. $4.50

NANTUCKET IN THE NINETEENTH CENTURY, Clay Lancaster. 180 rare photographs, stereographs, maps, drawings and floor plans recreate unique American island society. Authentic scenes of shipwreck, lighthouses, streets, homes are arranged in geographic sequence to provide walking-tour guide to old Nantucket existing today. Introduction, captions. 160pp. 8⅞ x 11¾.
23747-8 Pa. $6.95

STONE AND MAN: A PHOTOGRAPHIC EXPLORATION, Andreas Feininger. 106 photographs by *Life* photographer Feininger portray man's deep passion for stone through the ages. Stonehenge-like megaliths, fortified towns, sculpted marble and crumbling tenements show textures, beauties, fascination. 128pp. 9¼ x 10¾.
23756-7 Pa. $5.95

CIRCLES, A MATHEMATICAL VIEW, D. Pedoe. Fundamental aspects of college geometry, non-Euclidean geometry, and other branches of mathematics: representing circle by point. Poincare model, isoperimetric property, etc. Stimulating recreational reading. 66 figures. 96pp. 5⅝ x 8¼.
63698-4 Pa. $2.75

THE DISCOVERY OF NEPTUNE, Morton Grosser. Dramatic scientific history of the investigations leading up to the actual discovery of the eighth planet of our solar system. Lucid, well-researched book by well-known historian of science. 172pp. 5⅜ x 8½.
23726-5 Pa. $3.50

THE DEVIL'S DICTIONARY. Ambrose Bierce. Barbed, bitter, brilliant witticisms in the form of a dictionary. Best, most ferocious satire America has produced. 145pp. 5⅜ x 8½.
20487-1 Pa. $2.25

CATALOGUE OF DOVER BOOKS

HISTORY OF BACTERIOLOGY, William Bulloch. The only comprehensive history of bacteriology from the beginnings through the 19th century. Special emphasis is given to biography-Leeuwenhoek, etc. Brief accounts of 350 bacteriologists form a separate section. No clearer, fuller study, suitable to scientists and general readers, has yet been written. 52 illustrations. 448pp. 5⅝ x 8¼. 23761-3 Pa. $6.50

THE COMPLETE NONSENSE OF EDWARD LEAR, Edward Lear. All nonsense limericks, zany alphabets, Owl and Pussycat, songs, nonsense botany, etc., illustrated by Lear. Total of 321pp. 5⅜ x 8½. (Available in U.S. only) 20167-8 Pa. $3.95

INGENIOUS MATHEMATICAL PROBLEMS AND METHODS, Louis A. Graham. Sophisticated material from Graham *Dial*, applied and pure; stresses solution methods. Logic, number theory, networks, inversions, etc. 237pp. 5⅜ x 8½. 20545-2 Pa. $4.50

BEST MATHEMATICAL PUZZLES OF SAM LOYD, edited by Martin Gardner. Bizarre, original, whimsical puzzles by America's greatest puzzler. From fabulously rare *Cyclopedia*, including famous 14-15 puzzles, the Horse of a Different Color, 115 more. Elementary math. 150 illustrations. 167pp. 5⅜ x 8½. 20498-7 Pa. $2.75

THE BASIS OF COMBINATION IN CHESS, J. du Mont. Easy-to-follow, instructive book on elements of combination play, with chapters on each piece and every powerful combination team—two knights, bishop and knight, rook and bishop, etc. 250 diagrams. 218pp. 5⅜ x 8½. (Available in U.S. only) 23644-7 Pa. $3.50

MODERN CHESS STRATEGY, Ludek Pachman. The use of the queen, the active king, exchanges, pawn play, the center, weak squares, etc. Section on rook alone worth price of the book. Stress on the moderns. Often considered the most important book on strategy. 314pp. 5⅜ x 8½. 20290-9 Pa. $4.50

LASKER'S MANUAL OF CHESS, Dr. Emanuel Lasker. Great world champion offers very thorough coverage of all aspects of chess. Combinations, position play, openings, end game, aesthetics of chess, philosophy of struggle, much more. Filled with analyzed games. 390pp. 5⅜ x 8½. 20640-8 Pa. $5.00

500 MASTER GAMES OF CHESS, S. Tartakower, J. du Mont. Vast collection of great chess games from 1798-1938, with much material nowhere else readily available. Fully annotated, arranged by opening for easier study. 664pp. 5⅜ x 8½. 23208-5 Pa. $7.50

A GUIDE TO CHESS ENDINGS, Dr. Max Euwe, David Hooper. One of the finest modern works on chess endings. Thorough analysis of the most frequently encountered endings by former world champion. 331 examples, each with diagram. 248pp. 5⅜ x 8½. 23332-4 Pa. $3.75

CATALOGUE OF DOVER BOOKS

SECOND PIATIGORSKY CUP, edited by Isaac Kashdan. One of the greatest tournament books ever produced in the English language. All 90 games of the 1966 tournament, annotated by players, most annotated by both players. Features Petrosian, Spassky, Fischer, Larsen, six others. 228pp. 5⅜ x 8½. 23572-6 Pa. $3.50

ENCYCLOPEDIA OF CARD TRICKS, revised and edited by Jean Hugard. How to perform over 600 card tricks, devised by the world's greatest magicians: impromptus, spelling tricks, key cards, using special packs, much, much more. Additional chapter on card technique. 66 illustrations. 402pp. 5⅜ x 8½. (Available in U.S. only) 21252-1 Pa. $4.95

MAGIC: STAGE ILLUSIONS, SPECIAL EFFECTS AND TRICK PHOTOGRAPHY, Albert A. Hopkins, Henry R. Evans. One of the great classics; fullest, most authorative explanation of vanishing lady, levitations, scores of other great stage effects. Also small magic, automata, stunts. 446 illustrations. 556pp. 5⅜ x 8½. 23344-8 Pa. $6.95

THE SECRETS OF HOUDINI, J. C. Cannell. Classic study of Houdini's incredible magic, exposing closely-kept professional secrets and revealing, in general terms, the whole art of stage magic. 67 illustrations. 279pp. 5⅜ x 8½. 22913-0 Pa. $4.00

HOFFMANN'S MODERN MAGIC, Professor Hoffmann. One of the best, and best-known, magicians' manuals of the past century. Hundreds of tricks from card tricks and simple sleight of hand to elaborate illusions involving construction of complicated machinery. 332 illustrations. 563pp. 5⅜ x 8½. 23623-4 Pa. $6.00

MADAME PRUNIER'S FISH COOKERY BOOK, Mme. S. B. Prunier. More than 1000 recipes from world famous Prunier's of Paris and London, specially adapted here for American kitchen. Grilled tournedos with anchovy butter, Lobster a la Bordelaise, Prunier's prized desserts, more. Glossary. 340pp. 5⅜ x 8½. (Available in U.S. only) 22679-4 Pa. $3.00

FRENCH COUNTRY COOKING FOR AMERICANS, Louis Diat. 500 easy-to-make, authentic provincial recipes compiled by former head chef at New York's Fitz-Carlton Hotel: onion soup, lamb stew, potato pie, more. 309pp. 5⅜ x 8½. 23665-X Pa. $3.95

SAUCES, FRENCH AND FAMOUS, Louis Diat. Complete book gives over 200 specific recipes: bechamel, Bordelaise, hollandaise, Cumberland, apricot, etc. Author was one of this century's finest chefs, originator of vichyssoise and many other dishes. Index. 156pp. 5⅜ x 8.
23663-3 Pa. $2.75

TOLL HOUSE TRIED AND TRUE RECIPES, Ruth Graves Wakefield. Authentic recipes from the famous Mass. restaurant: popovers, veal and ham loaf, Toll House baked beans, chocolate cake crumb pudding, much more. Many helpful hints. Nearly 700 recipes. Index. 376pp. 5⅜ x 8½.
23560-2 Pa. $4.50

CATALOGUE OF DOVER BOOKS

THE CURVES OF LIFE, Theodore A. Cook. Examination of shells, leaves, horns, human body, art, etc., in *"the* classic reference on how the golden ratio applies to spirals and helices in nature . . . . "—Martin Gardner. 426 illustrations. Total of 512pp. 5⅜ x 8½. 23701-X Pa. $5.95

AN ILLUSTRATED FLORA OF THE NORTHERN UNITED STATES AND CANADA, Nathaniel L. Britton, Addison Brown. Encyclopedic work covers 4666 species, ferns on up. Everything. Full botanical information, illustration for each. This earlier edition is preferred by many to more recent revisions. 1913 edition. Over 4000 illustrations, total of 2087pp. 6⅛ x 9¼. 22642-5, 22643-3, 22644-1 Pa., Three-vol. set $25.50

MANUAL OF THE GRASSES OF THE UNITED STATES, A. S. Hitchcock, U.S. Dept. of Agriculture. The basic study of American grasses, both indigenous and escapes, cultivated and wild. Over 1400 species. Full descriptions, information. Over 1100 maps, illustrations. Total of 1051pp. 5⅜ x 8½. 22717-0, 22718-9 Pa., Two-vol. set $15.00

THE CACTACEAE,, Nathaniel L. Britton, John N. Rose. Exhaustive, definitive. Every cactus in the world. Full botanical descriptions. Thorough statement of nomenclatures, habitat, detailed finding keys. The one book needed by every cactus enthusiast. Over 1275 illustrations. Total of 1080pp. 8 x 10¼. 21191-6, 21192-4 Clothbd., Two-vol. set $35.00

AMERICAN MEDICINAL PLANTS, Charles F. Millspaugh. Full descriptions, 180 plants covered: history; physical description; methods of preparation with all chemical constituents extracted; all claimed curative or adverse effects. 180 full-page plates. Classification table. 804pp. 6½ x 9¼. 23034-1 Pa. $12.95

A MODERN HERBAL, Margaret Grieve. Much the fullest, most exact, most useful compilation of herbal material. Gigantic alphabetical encyclopedia, from aconite to zedoary, gives botanical information, medical properties, folklore, economic uses, and much else. Indispensable to serious reader. 161 illustrations. 888pp. 6½ x 9¼. (Available in U.S. only) 22798-7, 22799-5 Pa., Two-vol. set $13.00

THE HERBAL or GENERAL HISTORY OF PLANTS, John Gerard. The 1633 edition revised and enlarged by Thomas Johnson. Containing almost 2850 plant descriptions and 2705 superb illustrations, Gerard's *Herbal* is a monumental work, the book all modern English herbals are derived from, the one herbal every serious enthusiast should have in its entirety. Original editions are worth perhaps $750. 1678pp. 8½ x 12¼. 23147-X Clothbd. $50.00

MANUAL OF THE TREES OF NORTH AMERICA, Charles S. Sargent. The basic survey of every native tree and tree-like shrub, 717 species in all. Extremely full descriptions, information on habitat, growth, locales, economics, etc. Necessary to every serious tree lover. Over 100 finding keys. 783 illustrations. Total of 986pp. 5⅜ x 8½. 20277-1, 20278-X Pa., Two-vol. set $11.00

CATALOGUE OF DOVER BOOKS

AMERICAN BIRD ENGRAVINGS, Alexander Wilson et al. All 76 plates. from Wilson's *American Ornithology* (1808-14), most important ornithological work before Audubon, plus 27 plates from the supplement (1825-33) by Charles Bonaparte. Over 250 birds portrayed. 8 plates also reproduced in full color. 111pp. 9⅜ x 12½. 23195-X Pa. $6.00

CRUICKSHANK'S PHOTOGRAPHS OF BIRDS OF AMERICA, Allan D. Cruickshank. Great ornithologist, photographer presents 177 closeups, groupings, panoramas, flightings, etc., of about 150 different birds. Expanded *Wings in the Wilderness*. Introduction by Helen G. Cruickshank. 191pp. 8¼ x 11. 23497-5 Pa. $6.00

AMERICAN WILDLIFE AND PLANTS, A. C. Martin, et al. Describes food habits of more than 1000 species of mammals, birds, fish. Special treatment of important food plants. Over 300 illustrations. 500pp. 5⅜ x 8½. 20793-5 Pa. $4.95

THE PEOPLE CALLED SHAKERS, Edward D. Andrews. Lifetime of research, definitive study of Shakers: origins, beliefs, practices, dances, social organization, furniture and crafts, impact on 19th-century USA, present heritage. Indispensable to student of American history, collector. 33 illustrations. 351pp. 5⅜ x 8½. 21081-2 Pa. $4.50

OLD NEW YORK IN EARLY PHOTOGRAPHS, Mary Black. New York City as it was in 1853-1901, through 196 wonderful photographs from N.-Y. Historical Society. Great Blizzard, Lincoln's funeral procession, great buildings. 228pp. 9 x 12. 22907-6 Pa. $8.95

MR. LINCOLN'S CAMERA MAN: MATHEW BRADY, Roy Meredith. Over 300 Brady photos reproduced directly from original negatives, photos. Jackson, Webster, Grant, Lee, Carnegie, Barnum; Lincoln; Battle Smoke, Death of Rebel Sniper, Atlanta Just After Capture. Lively commentary. 368pp. 8⅜ x 11¼. 23021-X Pa. $8.95

TRAVELS OF WILLIAM BARTRAM, William Bartram. From 1773-8, Bartram explored Northern Florida, Georgia, Carolinas, and reported on wild life, plants, Indians, early settlers. Basic account for period, entertaining reading. Edited by Mark Van Doren. 13 illustrations. 141pp. 5⅜ x 8½. 20013-2 Pa. $5.00

THE GENTLEMAN AND CABINET MAKER'S DIRECTOR, Thomas Chippendale. Full reprint, 1762 style book, most influential of all time; chairs, tables, sofas, mirrors, cabinets, etc. 200 plates, plus 24 photographs of surviving pieces. 249pp. 9⅞ x 12¾. 21601-2 Pa. $7.95

AMERICAN CARRIAGES, SLEIGHS, SULKIES AND CARTS, edited by Don H. Berkebile. 168 Victorian illustrations from catalogues, trade journals, fully captioned. Useful for artists. Author is Assoc. Curator, Div. of Transportation of Smithsonian Institution. 168pp. 8½ x 9½. 23328-6 Pa. $5.00

CATALOGUE OF DOVER BOOKS

THE EARLY WORK OF AUBREY BEARDSLEY, Aubrey Beardsley. 157 plates, 2 in color: *Manon Lescaut, Madame Bovary, Morte Darthur, Salome,* other. Introduction by H. Marillier. 182pp. 8⅛ x 11. 21816-3 Pa. $4.50

THE LATER WORK OF AUBREY BEARDSLEY, Aubrey Beardsley. Exotic masterpieces of full maturity: *Venus and Tannhauser, Lysistrata, Rape of the Lock, Volpone,* Savoy material, etc. 174 plates, 2 in color. 186pp. 8⅛ x 11. 21817-1 Pa. $5.95

THOMAS NAST'S CHRISTMAS DRAWINGS, Thomas Nast. Almost all Christmas drawings by creator of image of Santa Claus as we know it, and one of America's foremost illustrators and political cartoonists. 66 illustrations. 3 illustrations in color on covers. 96pp. 8⅜ x 11¼. 23660-9 Pa. $3.50

THE DORÉ ILLUSTRATIONS FOR DANTE'S DIVINE COMEDY, Gustave Doré. All 135 plates from Inferno, Purgatory, Paradise; fantastic tortures, infernal landscapes, celestial wonders. Each plate with appropriate (translated) verses. 141pp. 9 x 12. 23231-X Pa. $4.50

DORÉ'S ILLUSTRATIONS FOR RABELAIS, Gustave Doré. 252 striking illustrations of *Gargantua and Pantagruel* books by foremost 19th-century illustrator. Including 60 plates, 192 delightful smaller illustrations. 153pp. 9 x 12. 23656-0 Pa. $5.00

LONDON: A PILGRIMAGE, Gustave Doré, Blanchard Jerrold. Squalor, riches, misery, beauty of mid-Victorian metropolis; 55 wonderful plates, 125 other illustrations, full social, cultural text by Jerrold. 191pp. of text. 9⅜ x 12¼. 22306-X Pa. $7.00

THE RIME OF THE ANCIENT MARINER, Gustave Doré, S. T. Coleridge. Dore's finest work, 34 plates capture moods, subtleties of poem. Full text. Introduction by Millicent Rose. 77pp. 9¼ x 12. 22305-1 Pa. $3.50

THE DORE BIBLE ILLUSTRATIONS, Gustave Doré. All wonderful, detailed plates: Adam and Eve, Flood, Babylon, Life of Jesus, etc. Brief King James text with each plate. Introduction by Millicent Rose. 241 plates. 241pp. 9 x 12. 23004-X Pa. $6.00

THE COMPLETE ENGRAVINGS, ETCHINGS AND DRYPOINTS OF ALBRECHT DURER. "Knight, Death and Devil"; "Melencolia," and more—all Dürer's known works in all three media, including 6 works formerly attributed to him. 120 plates. 235pp. 8⅜ x 11¼. 22851-7 Pa. $6.50

MECHANICK EXERCISES ON THE WHOLE ART OF PRINTING, Joseph Moxon. First complete book (1683-4) ever written about typography, a compendium of everything known about printing at the latter part of 17th century. Reprint of 2nd (1962) Oxford Univ. Press edition. 74 illustrations. Total of 550pp. 6⅛ x 9¼. 23617-X Pa. $7.95

CATALOGUE OF DOVER BOOKS

THE COMPLETE WOODCUTS OF ALBRECHT DURER, edited by Dr. W. Kurth. 346 in all: "Old Testament," "St. Jerome," "Passion," "Life of Virgin," Apocalypse," many others. Introduction by Campbell Dodgson. 285pp. 8½ x 12¼. 21097-9 Pa. $7.50

DRAWINGS OF ALBRECHT DURER, edited by Heinrich Wolfflin. 81 plates show development from youth to full style. Many favorites; many new. Introduction by Alfred Werner. 96pp. 8⅛ x 11. 22352-3 Pa. $5.00

THE HUMAN FIGURE, Albrecht Dürer. Experiments in various techniques—stereometric, progressive proportional, and others. Also life studies that rank among finest ever done. Complete reprinting of *Dresden Sketchbook*. 170 plates. 355pp. 8⅜ x 11¼. 21042-1 Pa. $7.95

OF THE JUST SHAPING OF LETTERS, Albrecht Dürer. Renaissance artist explains design of Roman majuscules by geometry, also Gothic lower and capitals. Grolier Club edition. 43pp. 7⅞ x 10¾ 21306-4 Pa. $3.00

TEN BOOKS ON ARCHITECTURE, Vitruvius. The most important book ever written on architecture. Early Roman aesthetics, technology, classical orders, site selection, all other aspects. Stands behind everything since. Morgan translation. 331pp. 5⅜ x 8½. 20645-9 Pa. $4.50

THE FOUR BOOKS OF ARCHITECTURE, Andrea Palladio. 16th-century classic responsible for Palladian movement and style. Covers classical architectural remains, Renaissance revivals, classical orders, etc. 1738 Ware English edition. Introduction by A. Placzek. 216 plates. 110pp. of text. 9½ x 12¾. 21308-0 Pa. $10.00

HORIZONS, Norman Bel Geddes. Great industrialist stage designer, "father of streamlining," on application of aesthetics to transportation, amusement, architecture, etc. 1932 prophetic account; function, theory, specific projects. 222 illustrations. 312pp. 7⅞ x 10¾. 23514-9 Pa. $6.95

FRANK LLOYD WRIGHT'S FALLINGWATER, Donald Hoffmann. Full, illustrated story of conception and building of Wright's masterwork at Bear Run, Pa. 100 photographs of site, construction, and details of completed structure. 112pp. 9¼ x 10. 23671-4 Pa. $5.50

THE ELEMENTS OF DRAWING, John Ruskin. Timeless classic by great Viltorian; starts with basic ideas, works through more difficult. Many practical exercises. 48 illustrations. Introduction by Lawrence Campbell. 228pp. 5⅜ x 8½. 22730-8 Pa. $3.75

GIST OF ART, John Sloan. Greatest modern American teacher, Art Students League, offers innumerable hints, instructions, guided comments to help you in painting. Not a formal course. 46 illustrations. Introduction by Helen Sloan. 200pp. 5⅜ x 8½. 23435-5 Pa. $4.00

CATALOGUE OF DOVER BOOKS

GEOMETRY, RELATIVITY AND THE FOURTH DIMENSION, Rudolf Rucker. Exposition of fourth dimension, means of visualization, concepts of relativity as Flatland characters continue adventures. Popular, easily followed yet accurate, profound. 141 illustrations. 133pp. 5⅜ x 8½.
23400-2 Pa. $2.75

THE ORIGIN OF LIFE, A. I. Oparin. Modern classic in biochemistry, the first rigorous examination of possible evolution of life from nitrocarbon compounds. Non-technical, easily followed. Total of 295pp. 5⅜ x 8½.
60213-3 Pa. $4.00

PLANETS, STARS AND GALAXIES, A. E. Fanning. Comprehensive introductory survey: the sun, solar system, stars, galaxies, universe, cosmology; quasars, radio stars, etc. 24pp. of photographs. 189pp. 5⅜ x 8½. (Available in U.S. only)
21680-2 Pa. $3.75

THE THIRTEEN BOOKS OF EUCLID'S ELEMENTS, translated with introduction and commentary by Sir Thomas L. Heath. Definitive edition. Textual and linguistic notes, mathematical analysis, 2500 years of critical commentary. Do not confuse with abridged school editions. Total of 1414pp. 5⅜ x 8½. 60088-2, 60089-0, 60090-4 Pa., Three-vol. set $18.50

Prices subject to change without notice.

Available at your book dealer or write for free catalogue to Dept. GI, Dover Publications, Inc., 180 Varick St., N.Y., N.Y. 10014. Dover publishes more than 175 books each year on science, elementary and advanced mathematics, biology, music, art, literary history, social sciences and other areas.